The Evolution of Human Wisdom

The Evolution of Human Wisdom

Edited by Celia Deane-Drummond
and Agustín Fuentes

LEXINGTON BOOKS
Lanham • Boulder • New York • London

Published by Lexington Books
An imprint of The Rowman & Littlefield Publishing Group, Inc.
4501 Forbes Boulevard, Suite 200, Lanham, Maryland 20706
www.rowman.com

Unit A, Whitacre Mews, 26-34 Stannary Street, London SE11 4AB

British Library Cataloguing in Publication Information Available

Library of Congress Cataloging-in-Publication Data Available

ISBN 978-1-4985-4845-8 (cloth : alk. paper)
ISBN 978-1-4985-4846-5 (electronic)

∞™ The paper used in this publication meets the minimum requirements of American National Standard for Information Sciences Permanence of Paper for Printed Library Materials, ANSI/NISO Z39.48-1992.

Printed in the United States of America

Contents

Foreword

Jeremy Kendal

It is my pleasure to introduce this volume of contributions to an interdisciplinary dialogue between anthropologists and theologians on the evolution of the sapient qualities by which Carl Linnaeus defined our species. The proceeding contributions represent part of an ongoing interdisciplinary dialog across these divergent branches of thought. In engaging with them, the reader will be challenged to consider a diversity of concepts from the socio-cognitive human niche to the realms of angels. And, as is common upon meetings of diverged epistemological frameworks, some cross-disciplinary conceptualizations may contradict and not all hybrid ideas will be fertile. Nonetheless, this cross-disciplinary collection is likely not only to provide challenging reflection for those willing to venture from the safe confines of their home disciplines, but also to provoke genuine insight of an interdisciplinary nature.

The remarkable contributions of Darwin and Wallace contributed to schisms in Western philosophy, into disciplines including evolutionary anthropology, socio-cultural anthropology, and theology. Yet even contemporary Western commentaries on wisdom, and particularly Catholic theology, are still permeated with the writings of the thirteenth-century monk Thomas Aquinas, and as such, he plays a prominent role in this volume. If, as I, you were no expert in Thomism, you may yet have come across the following entertaining quote by G.K. Chesterton:

> Since the modern world began in the sixteenth century, nobody's system of philosophy has really corresponded to everybody's sense of reality; to what, if left to themselves, common men would call common sense. Each started with a paradox; a peculiar point of view demanding the sacrifice of what they would call a sane point of view. That is the one thing common to Hobbes and Hegel, to Kant and Bergson, to Berkeley and William James. A man had to believe something that no normal man would believe, if it were suddenly propounded to his simplicity; as that law is above right, or right is outside reason, or things are only as we think them, or everything is relative to a reality that is not there. The modern philosopher claims, like a sort of confidence man, that if we will grant him this, the rest will be easy; he will straighten out the world, if he is allowed to give this one twist to the mind . . . Against all this the philosophy of St. Thomas stands founded

> on the universal common conviction that eggs are eggs. The Hegelian may say that an egg is really a hen, because it is a part of an endless process of Becoming; the Berkelian may hold that poached eggs only exist as a dream exists, since it is quite as easy to call the dream the cause of the eggs as the eggs the cause of the dream; the Pragmatist may believe that we get the best out of scrambled eggs by forgetting that they ever were eggs, and only remembering the scramble. But no pupil of St. Thomas needs to addle his brains in order adequately to addle his eggs; to put his head at any peculiar angle in looking at eggs, or squinting at eggs, or winking the other eye in order to see a new simplification of eggs. The Thomist stands in the broad daylight of the brotherhood of men, in their common consciousness that eggs are not hens or dreams or mere practical assumptions; but things attested by the Authority of the Senses, which is from God.[1]

From an ancestral Aristotelian influence, his view of Thomism ends by highlighting an empirically derived approach to the understanding of things that is also fundamental to modern science. And in this volume also, much of the content concerns the qualities of wisdom evident in humans' response to, or in conjunction with, sense-derived data. Following Linnaeus and Aquinas, we might consider that wisdom, by definition, pertains to those distinctive aspects of human intellect and ways of being that are not found in other animals. In this spirit, the volume offers definitions of wisdom that range from aspects of practical wisdom, often aligned with the Cardinal Virtue, prudence, to the realization of an eternal essence, akin perhaps to "logos" as used by St. Augustine and in Sufism, or a state of emptiness, indicated by "padme" in the Tibetan Buddhist mantra "Om mani padme hum."

Before examining what is unique about humans, it is useful to contextualize by considering attributes that we share with other species. Many fauna exhibit phenotypic flexibility, adaptively responding to sensory inputs from their environment. Furthermore, the challenge of surviving an environment that changes within an individual's lifetime has resulted in the evolution of the capacity to learn. In many cases, associational learning of temporally or spatially correlated events in the environment, or of the relationship between a stimulus and the outcome of a particular behavior, is an effective mechanism to find food, avoid predation, or affect social rank. There are also many examples of non-human animals learning from one another in capricious environments, often using heuristics guiding when to engage in the low-cost scrounging of information from others and whom to copy in order to ameliorate the risks that acquired knowledge may be outdated or ineffectual.[2]

In modern humans, associational learning is commonplace. Pavlov's dog experiment would work just as well with humans as with dogs, albeit perhaps with donuts or cookies in place of dog food! And of course we are excellent at social learning, but surely human cognition in the wild is more complicated than this? After all, we are capable of abstraction, reasoning, induction, reflexivity, hermeneutics, etc. While the non-human animal's *umwelt* might be saturated by their immediately perceived environment,[3] ours is refracted by the lens of culturally specified ontology, translating our sensory input into something that shapes, and is shaped by, meaning. As humans reproduce and die, so ontologies persist and undergo modification across generations by continual reconstruction through legitimate peripheral participation[4] in a local community ecology, consisting of human, non-human, biotic and abiotic relations. Humans have also exhibited the cumulative cultural evolution of technologies and skills that no one individual could hope to invent alone. Thus we suspect that humans are capable of unusual cognition such as self-conscious reflection, imagination, future-thinking, shared intentionality, theory of mind, symbolic thought, language, etc. But this heady mix is situated in the life of *habitus*[5] for which simple conditioning and learning heuristics may contribute to the development of normative practice.

A theme that some readers may find resonating through the volume is the relationship between forms of detachment and non-independence from environment. In what follows, I cite contributing authors simply to highlight where the reader can find discussion around a particular topic, but without intention to recapitulate or pigeonhole their arguments here. Aquinas asserted that humans, unlike non-human animals, possess intelligence (see Willows). Fast forward over 800 years to modern evolutionary theory, and a prominent reassessment of this claim is that humans constructed a cultural niche (see Fuentes) that co-evolved with a suite of socio-cognitive processes including future thinking (see Deane-Drummond), theory-of-mind (see Baynes-Rock), symbolic thought (see Kissel), speculation (see Belton), and lying (Clem). Typically, these cognitive processes are thought to involve some form of abstraction, or representation, that implies detachment from the environment.

Yet simultaneously, processes such as ostensive inferential communication (see Fuentes) facilitate the construction of meaning by developing shared or collective intentionality (see Iffland). Here, detachment of the individual from their social environment is dissolved both in the process by which meaning is generated and in the semantic qualities of meaning. Similarly, homemaking (see Feder) requires practices that interact with a specific place to affect personal and social identities that are non-independent of that place. These procedural and ontological interactions with the environment can be extended further to include dynamic human relations with other species (see Campbell). Thus, humans may have evolved the cognitive capacity for metacognition and abstract representation that is, by definition, detached from the immediate environment, yet the empirical processes of generating meaning and the concomitant qualities of constructed ontology are, by definition, non-independent of environment.

If there is a lesson to take from this dichotomous relationship with the environment, it is actually one that is already made explicit in the notion of the cultural niche, namely that wisdom needs to be studied in relation to the partially-constructed environment within which it is practiced.[6] In these circumstances, cognition can be distributed and mediated by artifacts and normative practices affecting structured learning environments.[7] Wise synthetic thought processes such as reasoning or mind reading may be shaped by the specific cultural environment in which humans develop[8] and may to some degree develop in the unconscious mind.[9] It may be that conscious experience, forming the perception of a pre-reflexive self-consciousness,[10] facilitates rationalization and the sharing of experience[11] to affect "wise" practice and construction of meaning in the human cultural niche. And with that, I invite the reader to engage in the broad, thought-provoking discussion to be found in the chapters ahead.

NOTES

1. G. K. Chesterton, *St. Thomas Aquinas* (Mineola, NY: Dover Publications, 1933), 147.
2. Kevin N. Laland, "Social Learning Strategies," *Animal Learning and Behavior* 32.1 (2004): 4–14.
3. Kalevi Kull, "On Semiosis, Umwelt, and Semiosphere," *Semiotica* 120 (1998): 299–310.
4. Jean Lave and Étienne Wenger, *Situated Learning: Legitimate Peripheral Participation* (Cambridge: Cambridge University Press, 1991).
5. Pierre Bordieu, *The Logic of Practice* (Cambridge: Polity Press, 1990).
6. Edwin Hutchins, *Cognition in the Wild* (Cambridge, MA: MIT Press, 1995).
7. Kim Sterelny, *The Evolved Apprentice: How Evolution Made Humans Unique* (Cambridge, MA: MIT Press, 2012).
8. Cecilia M. Heyes and Chris D. Frith, "The Cultural Evolution of Mind Reading," *Science* 344.6190 (2014): 1243091.
9. Hugo Mercier and Dan Sperber, *The Enigma of Reason* (Harvard: Harvard University Press, 2017).
10. Dorothée Legrand, "Pre-Reflective Self-as-Subject from Experiential and Empirical Perspectives," *Consciousness and Cognition* 16 (2007): 583–99.
11. Chris D. Frith, "Consciousness Is for Sharing," *Cognitive Neuroscience* 2.2 (2011): 117–18; Mercier and Sperber, *Enigma of Reason*.

BIBLIOGRAPHY

Bourdieu, Pierre. *The Logic of Practice*. Cambridge: Polity Press, 1990.

Chesterton, G. K. *St. Thomas Aquinas*. Mineola, NY: Dover Publications, 1933.

Frith, Chris D. "Consciousness Is for Sharing." *Cognitive Neuroscience* 2.2 (2011): 117–18.

Heyes, Cecilia M., and Chris D. Frith. "The Cultural Evolution of Mind Reading." *Science* 344.6190 (2014): 1243091.

Hutchins, Edwin. *Cognition in the Wild*. Cambridge, MA: MIT Press, 1995.

Kull, Kalevi. "On Semiosis, Umwelt, and Semiosphere." *Semiotica* 120 (1998): 299–310.

Laland, Kevin N. "Social Learning Strategies." *Animal Learning and Behavior* 32.1 (2004): 4–14.

Lave, Jean, and Étienne Wenger. *Situated Learning: Legitimate Peripheral Participation*. Cambridge: Cambridge University Press, 1991.

Legrand, Dorothée. "Pre-Reflective Self-as-Subject from Experiential and Empirical Perspectives." *Consciousness and Cognition* 16 (2007): 583–99.

Mercier, Hugo, and Dan Sperber. *The Enigma of Reason*. Harvard: Harvard University Press, 2017.

Sterelny, Kim. *The Evolved Apprentice: How Evolution Made Humans Unique*. Cambridge, MA: MIT Press, 2012.

Acknowledgments

Seeds for the ideas in this book were first discussed during a research project hosted by the Center for Theological Inquiry in Princeton, where both editors were privileged to spend a year from 2012–2013 researching the topic of human nature and evolution as part of a three-year CTI initiative funded by the John Templeton Foundation. The following year, in 2014, we were awarded a Notre Dame grant from the John Templeton Foundation (JTF) for a project entitled *The Evolution of Wisdom* and we were awarded a second JTF grant in 2015 for a larger project entitled *Human Distinctiveness*, whose special research focus was the evolution of wisdom. Arising out of the *Evolution of Wisdom* project, and in partnership with the *Human Distinctiveness* project, in July 2016 we held a research colloquium at the Institute of Advanced Study (IAS) at Durham University, and the chapters in this book are a direct result of the debates on working papers presented at that meeting. We are particularly grateful for the opportunities that the John Templeton Foundation has provided through funding these projects, to CTI for hosting a wonderful year of research, and to IAS (Durham) for giving us access to their facilities. We are also grateful to Robert Song and Robert Barton from Durham University's theology and anthropology departments respectively who took part in the discussions in addition to those from Durham University already listed in this volume. This book was also made possible in part by support in the form of an indexing and copyediting grant from the Institute for Scholarship in the Liberal Arts, College of Arts and Letters, University of Notre Dame. We are particularly grateful to Marilyn Burton for her expert copy-editing of the manuscript and preparation of the index, and to Rebecca Artinian-Kaiser, Assistant Director for the Center for Theology, Science and Human Flourishing, without whose support this book project would not have been realized.

This publication was made possible through the support of a grant from the John Templeton Foundation. The opinions expressed in this publication are those of the authors and do not necessarily reflect the views of the John Templeton Foundation.

Introduction

An Evolving Quest for Human Wisdom

Celia Deane-Drummond and Agustín Fuentes

This book addresses some of the most pressing and interesting questions about the puzzle of human origins and human distinctiveness related to a specific focus on the evolution of wisdom. How can we best understand the human capacity for wisdom, where did it come from, and how did it emerge? In the humanities and social sciences, wisdom is often said to be a social and cognitive capacity that serves to ensure the quality of community function. More specifically, wisdom in an evolutionary anthropological approach can be seen as the capacity to create and manipulate material items and social relations with increasing skill and imagination. In this context, the human capacity and deployment of wisdom is part and parcel of what some scholars have termed the "human niche."[1] In theology, wisdom is defined more broadly and may be associated with the task of theology as such, namely, a second order interpretation of human social life through images and concepts based on human religious experience. Wisdom is also integral to a particular style of biblical literature that is more open to insights from other cultural traditions. Biblical wisdom is never closed in that it implies a search for knowing that is both alluring and never finished. Such wisdom thinking includes not just practical wisdom in relation to efficacy for action in the material world, but also more transcendent forms of wisdom that are even harder to define. In Christian theology, Christ comes to represent incarnate divine Wisdom in the world. What has not yet been attempted in a rigorous way is an engagement between theology and the sciences, especially insofar as the evolutionary science develops the parameters for precisely how the capacity for different facets of human wisdom might have evolved. What are the basic parameters, and processes, of the evolutionary origins of human wisdom? For the pragmatic purposes of developing an integrated approach, we can take general anthropological definitions of wisdom as the pattern of (and capacity for) successful complex decision making in navigating social networks and dynamic niches in human communities. Such a definition is barely sufficient for theology, but at least it means that the conversation between theologians and anthropologists can begin. Certainly, theological wisdom would presuppose this anthropological envisioning of wisdom as a necessary backdrop to developing higher forms of wisdom. Using wisdom as a way of framing such discussion shifts the standard paradigms in evolutionary anthropology in significant ways. Furthermore, insights from evolutionary research raise new and interesting questions about human origins for philosophy and theology.

This book quite deliberately brings together both established and emerging scholars in the fields of theology and anthropology. In doing so, we find lines of convergence and divergence between Christian theology and evolutionary anthropology in defining and exploring different aspects of wisdom. Most notable in this debate are the following: first, what are the philosophical difficulties that arise when two very different methodological approaches to the manner of humans becoming wise are brought together in discussion? Second, what is the relative importance and significance of human language in defining the meaning of wisdom and its expression? Third, to what extent does a theologically informed wisdom discourse push evolutionary anthropology to formulate new questions and *vice versa*? While, on the one hand, anthropologists search for signs, signals, and implications that may reflect actions emerging from human wisdom in the human evolutionary record, theologians, on the other hand, ask how the integration of theological insights might shape, facilitate, and change the ways these questions about human wisdom are answered. Furthermore, the evolutionary anthropological perspective on wisdom raises important questions for theological anthropology to address—for example, what might it mean to find evidence of traces of human wisdom far deeper in the human lineage than originally anticipated? This book shows, however, drawing on philosophical analysis, that there is no simple consonance between evolutionary anthropology and theology. Yet, at the same time, each discipline has much to learn from the other; even in the midst of an awareness of dissonance and some tension, there can still be mutual respect. The goal of this book is to begin to develop a trans-disciplinary approach to the evolution of human wisdom, where each discipline is challenged to ask questions in a new way as a result of engagement with another field of inquiry.

The book begins with a prologue that lays out the Philosophical Parameters for our discussion. What kinds of explanations are appropriate for us to make sense of the phenomenon of wisdom from an evolutionary perspective? In a presentation that formed the initial discussion for this book, Robert Song addressed the issue from the perspective of the largest possible philosophical frame of reference, namely, the metaphysical conditions for the possibility of human wisdom.[2] Starting with a critical analysis of Thomas Nagel's widely-discussed *Mind and Cosmos*,[3] Song endorsed Nagel's conclusion that naturalist accounts of evolution are fundamentally incapable of explaining the emergence of consciousness, cognition, and value. He argued, however, that Nagel's own solution of a non-theistic teleology fails for several reasons. First, it is itself explanatorily inadequate, not least in relation to the question of why anything exists at all. Second, Nagel's approach fundamentally misunderstands the nature of the doctrine of creation as classically theologically conceived where God is the ground of all creation and at the same time fundamentally distinct from it. Third, it conceives the philosophical task as one that will specifically enable explanatory/predictive success. This fails to grasp the fundamental nature of knowledge as related to humanity's encounter with the world, which is central to the theological task as well. It fails, that is, to grasp the nature of philosophy as integrally (as well as etymologically) related to both wisdom and love.

Song argued that a fundamentally alternative explanation of wisdom can be addressed through metaphor theory, which notes that wisdom involves dimensions of knowledge that are not just cognitive-discursive but also involve appropriate response and behavior, thereby integrating intellectual and moral virtues. In particular, wisdom is always predicated on reality that is ultimately personal. On this account, human agency, far from being an embarrassing phenomenon that needs further explanation, an *explanandum*, becomes the epistemic center of all explanation. Thus knowledge of persons, which involves both wisdom and love, becomes the aboriginal form of all knowledge. Set in a theological context, this suggests that the evolution of wisdom is ultimately only possible not because wisdom "emerges" from something lower than itself, but because it is a reflection of the wisdom that pervades the universe. Clearly, this is a very different approach to wisdom compared with that possible through evolutionary understandings and analyses of archeological artifacts; yet, by presenting this challenge, Song reminds both science and secular philosophy of the limitations of different forms of knowledge.

In a slightly different vein, but touching on a similar frustration with reductionistic accounts of wisdom, Dylan Belton recovers Max Scheler's ideas in a chapter entitled "The Human as World-Open Spirit: An Exploration into Philosophical Anthropology and the Foundations of Human Wisdom." In this chapter, he seeks to introduce the work of the philosopher Max Scheler into reflections on human wisdom. Scheler was the founder of the twentieth-century German tradition of *Philosophische Anthropologie*, which, he argues, can serve as an important resource in the current interdisciplinary attempt to move beyond the neo-Darwinian paradigm. Within the latter, which tends to be dominated by a reductionistic and adaptationist logic, human intellectual capacities will always be understood instrumentally in terms of survival and reproduction and therefore will be explicated in terms of the development of principles already present within the rest of the animal kingdom. His use of Scheler attempts to challenge this vision and its underlying logic. Central to Scheler's philosophical anthropology is his notion of spirit. Scheler introduces spirit as a principle in order to explain at the metaphysical level what he takes to be the defining feature of the human being in comparison to the rest of the animal world, namely that of *world openness.* Becoming human is, therefore, according to Scheler, the process of elevation into world openness by virtue of spirit. He focuses on one specific act that Scheler deems to be fundamental to spirit and hence to the human being's world openness, the act known as *ideation*. This act is constitutive of what may be called human speculative wisdom, and Belton argues that it is qualitatively distinct from the acts that constitute the technical or practical intelligence that humans share with certain other animals.

The first section of this book, entitled Signs in Evolutionary Anthropology, begins with a chapter by Marc Kissel entitled "Piercing Peirce: Using Semiotics to Understand the Origins of Human Wisdom." According to Kissel, wisdom is somewhat illusory for anthropologists, and therefore not a topic that evolutionary anthropologists tend to address in a straightforward manner. He argues, however, that there are similarities in the ways in which evolutionary anthropologists seek to understand human evolutionary history. Contemporary debates in evolutionary anthropology have focused on questions such as how hominins gradually went from being more ape-like to more human-like, given the split from the common ancestor between apes and humans 8–10 million years ago. Such debates also almost always involve discussion of the origins of symbolic thought. In this chapter, he discusses some ways in which anthropologists have approached this question, concentrating on the fossil, genetic, and archeological data. While these data sets are salient for the question of human wisdom as defined from an anthropological perspective, he suggests that care must be taken in applying the material from the archeological and biological records to this question, as there are often assumptions in these models, relative to the broader understanding of what human wisdom might entail for evolutionary processes, that are untested or at least not fully developed. The complex answer to what makes humans "sapiens" no doubt entails not just anthropological data, but the incorporation of theories from neuroscience, information theory, semiotics, innovation theory, and genomics, as well as relevant discussions with the humanities. If wisdom can be seen as central to complex decision making when successfully navigating social network and dynamic niches, then its origins are most likely found in early, possibly pre-*Homo sapiens*, humans.

Marcus Baynes-Rock fills out this discussion with his commentary on "The Minds of Others," probing an important and largely un-researched question: what can the domestication of animals teach us about human wisdom? He argues that a key element of human wisdom as engaged within this book is dependent on a prior acceptance of *theory of mind,* understood as the ability to make layered inferences about the mental states of others. He suggests that theory of mind enabled ancestral humans to navigate a complex social niche replete with other minds, not just of other humans but of other animate beings. Ethnographic research shows that such a pattern persists not only in hunter/gatherer societies, but in agro-pastoral societies, and contrary to expectations. Baynes-Rock even asserts that such practices and patterns may still be present in societies tied to modern farming paradigms. The persistence of theory of mind as a basal mechanism in the human psyche is accentuated within the field of research into autism, a disorder that inhibits the human capacity for attributions of mental states to others, or the identifying of such states as similar to or the same as one's own. It appears from that research that theory of mind is inveterate in non-autistic humans and fosters the ascription of mental states not only to people and animals, but to animated shapes, weather phenomena, and—according to a metaphysical theory of mind—gods. Investigations into autism show that the capacity for theory of mind is not crucial to animal domestication. Baynes-Rock believes that his foray into animal domestication and the minds of others shows that human becoming was and is characterized by a diversity of paths for navigating the more-than-human world.

In the second section, entitled Evolving Homespun Wisdom, Ben Campbell's chapter on "Growing Wisdom" investigates later human archeological history in the Neolithic. He asks whether the use of ethnographic data and comparative anthropological analyses of farming communities can help to promote further discussion about imaginative emergence in relationships between humans and non-humans, and among humans in company with non-humans in landscapes of reciprocal attention. This chapter discusses ethnographic analyses arising from close observation of peoples growing food, in terms of symbolic and other forms of meaning, and the properties of relationships with animal and plant domesticates in such contexts. He argues that food species are not inert resources of material production; rather, they enliven domestic hearths and everyday social landscapes and are imbricated in everyday and ritualized practices of being social persons. Environmental personhood flows from gender-based divisions of activity and cultural expectation of gift and return. Social collectivities are brought into being and differences negotiated in culturing lifecycles of crops and livestock, and in maintaining conscious distinction from others' ways of culturing. Reflecting on how the Neolithic reinvention of relationships with species and landscapes has been rethought by different generations of scholarship, Campbell identifies some areas of theological concern over wisdoms of companiability, mastery, and sustainability.

Julia Feder's chapter "*Homo sapiens sapiens*: The Human as Homemaker" provides a companion to Campbell's more focused ethnographic study. Drawing on recent evolutionary theory, she takes human niche construction as a starting point for reflection on Christian wisdom as homemaking. In the first half of this essay, she argues that Christian theology needs to take account of the ways in which time and place are not mere impositions upon human beings. The influential twentieth-century Catholic theologian Edward Schillebeeckx seemed to assume this was the case by his inclusion of conditioning by time and space among his "anthropological constants." Based on niche construction theory she argues that time, space, and human beings interact in a dynamic process of "mutual mutability."[4] In the second half of the paper, she offers a portrait of Jesus as homemaker—viz., as one who imaginatively creates and facilitates a home within the material constraints of his social and historical location. As one who both embodies the fullness of humanity and incarnates divine Wisdom, she argues that this reading of Jesus is able to facilitate a robust Christian reading of human wisdom as homemaking, in other words, simultaneously living in tune with the way the world is and actively constructing it as the kind of place in which we desire to live.

The third section of this book, entitled The Wisdom of Speech, tackles important questions about the significance of language in human becoming. Opening the section, Stewart Clem engages with Thomistic thought in his chapter entitled "Speaking Truthfully: On the Peculiar Origins of Human Speech." Writing near the end of the fourth century, Saint Augustine of Hippo observed, "Indeed, nothing should be thought more peculiar to humankind than lies and falsehoods, which derive from humanity itself."[5] He could not have anticipated that, sixteen centuries later, developments in anthropology, psychology, and evolutionary theory would lend important insights to this simple observation. While nonhuman animals are capable of deception, humans, Clem argues, seem to be the only species that has the linguistic capacity to tell *lies*. An important corollary to this is the fact that human language depends on a general presumption of truthfulness, or mutual trust, in order to function at all. Some researchers have hypothesized that the practice of deceptive vocal utterances has in fact hindered other species from developing the skills required for language. Yet, humans somehow overcame this barrier and developed sophisticated speech practices such as narrative and displaced reference. Scholars in the social and biological sciences have theorized about the preconditions of human language while developing concepts such as interdependence, social cooperation, and joint intentionality. These insights have led some scholars in other disciplines—especially in philosophy and theology—to dismiss pre-modern conceptions about the "natural" function of language, assuming that new insights have rendered such claims meaningless. In this chapter, Clem suggests that a close examination of the role of deception in the evolution of human language—from perspectives in anthropology, comparative psychology, and the philosophy of language—actually gives us a better grasp of what it might mean to say, along with Thomas Aquinas, that lying is "unnatural" for human beings.

Craig Iffland explores speech acts in children in a second chapter in this section entitled "Precursors to Explanations of Action: Collective Intentionality and the Wisdom of Early Childhood." In this chapter, Iffland examines a particular social practice, namely, the practice of offering explanations of action. His interest in exploring this practice is that moral discourse presupposes not only the intelligibility of asking for and giving explanations of action, but also the capacity to inhabit a normative, social order that provides common standards for the rectitude of such explanations. He focuses critical attention on the claim that possession of theory of mind is an important achievement in the cognitive development of children, particularly in their ability to offer explanations of action. He argues that this ability is rooted in the child's capacity for collective intentionality, that is, cognition of how we relate to some particular object of attention. This capacity enables young children to acquire the art of navigating a social world that routinely demands that they explain themselves. A key manifestation of this capacity is young children's ability to forge and comprehend a shared, experiential background or social frame with interlocutors. By relating to particular objects of cognition within this shared frame, eventually the application of standards of evaluation and explanation of action—the rule of practical wisdom—becomes possible.

The fourth section, Evolving Wisdom as Virtue, begins with a chapter by Adam Willows entitled "Change and Constancy in the Nature of Wisdom Over Time." He argues that from a theological starting point, wisdom is a very heavily laden term. To ascribe (or deny) wisdom to a creature is not simply to make a comment about their ability for judgment. It is also to comment on their place in the created order, the nature of their soul, the particular kinds of goods they need, and the character of their relationship with God. Using the thought of Thomas Aquinas, this chapter explores the wider theological implications of ascribing wisdom to our evolutionary ancestors. Willows begins with a discussion of things common to all wise creatures. For Aquinas, saying that our ancestors were wise means assigning full human nature, including immortality, to those ancestors. Any ancestors without the capacity for wisdom, however, are not included in this. He discusses the way that Aquinas' theory of the soul necessitates a sharp distinction between human and non-human which seems to be at odds with the gradual trajectory of our evolutionary history. However, Willows insists that the two accounts are not necessarily incompatible. He moves from discussing commonalities to the differences that may exist between wise creatures. According to Willows' interpretation, Aquinas does recognize that there may be differences in levels of wisdom between individuals, but does not think that this is exclusively determined by intellectual capacity. This means that saying our ancestors were wise involves accepting the possibility that they could have been wiser than modern humans.

In the second chapter in this section, entitled "Practical Wisdom in the Making: A Theological Approach to Early Hominin Evolution in Conversation with Modern Jewish Philosophy," Celia Deane-Drummond explores Jewish philosophical reflections on human identity and the specific way in which elements of practical wisdom or prudence may have evolved, and in which they show up in the evolutionary record. Practical wisdom is viewed in her account as a critical aspect of moral evolution in the *Homo* lineage. Trying to understand how novelty has appeared in evolutionary terms is often described through the philosophical language of emergence. Emergence narratives can sometimes create a false impression of a coalescent, progressive trajectory in the evolution of complexity that fails to give proper attention to the sporadic nature of such processes and fails to address non-emergent aspects of wisdom's appearance. The scattered occurrence of what looks like "symbolic" behavior of early hominins challenges an all-or-nothing appearance. In order to navigate this complex and difficult theological problem, she draws on the work of three modern Jewish philosophers—Alan Mittleman, Joseph Soloveitchik, and Hans Jonas—who in different ways have all engaged with both Darwinian evolutionary theory and the biblical account of human origins. In common with Willows, her suggestion is that a Thomistic account of practical wisdom is worth considering further, since its definition in terms of different parts of wisdom allows, in theory, for partial or incomplete aspects of practical wisdom appearing or even disappearing in evolutionary terms. From a theological perspective, there remain challenging issues to consider related to the extent to which divine immanence and transcendence are brought to bear in the creation of novelty in the early evolution of humans.

The final chapter in this collection gathers many different elements that have been discussed in a culminating Epilogue that seeks to address Questions and Puzzles in Evolutionary Anthropology. Agustín Fuentes' contribution, entitled "Manipulating Materials, Bodies, and Signs: The Ecology of Creative Problem Solving and Imaginative Sociality Set the Context for Language in the Later Pleistocene Human Niche," recognizes that language is critical to the contemporary human niche. He argues that key aspects of human evolution over the past fifteen to twenty millennia are tied to linguistic capacities and manipulation (e.g., gender, political, and economic hierarchies, expansive kinship structures, group identities, and sodalities, broad-scale domestication, inter-societal conflict and peace, etc.). But he suggests that the evolution of language, its first appearances, and anything that we might term "proto-language" are all areas where hypothesis-driven scientific analysis is extremely difficult. Early linguistic behavior is extremely difficult to detect in the archeological record; fossil data, and genetic and physiological assessments (contemporary and ancient), rarely provide specific insights that prove explanatory for the emergence of language. He suggests that there are other options for understanding the emergence of language *per se* in Pleistocene *Homo*. Rather than attempting to pinpoint or explain the emergence of language, he proposes that an approach examining the development of a human wisdom—the capacity to create and manipulate material items and social relations with increasing skill and imagination—might inform our models of the human niche in a manner that facilitates a better understanding of the contexts in which language could emerge. This chapter does not attempt to describe the evolution of language, or proto-language; rather, it seeks to describe and interpret aspects of the human niche in the later Pleistocene that provided the social, ecological, and neurobiological contexts in which language could develop. His core suggestion is that the emergence of a human wisdom may have been critical in facilitating the eventual emergence of human language.

We invite the reader, therefore, to explore questions about the evolution of human wisdom with the writers of this volume. We close the discussion by showing how the lens of wisdom opens up even more unsolved puzzles in evolutionary anthropology. A theologian reading such a text will be very familiar with the idea of wisdom as a collective search. The difference, in this book, is that we have companions in that search who are approaching that topic through a scientific lens, and thus opening up further questions for theological reflection, even if, in the end, the epistemological basis for that knowing remains distinct according to disciplinary commitments.

NOTES

1. See, for example, Agustín Fuentes, "Integrative Anthropology and the Human Niche: Toward a Contemporary Approach to Human Evolution," *American Anthropologist* 117 (2015): 302–15.
2. Unfortunately, Robert Song was unable to contribute a chapter to this volume due to unforeseen personal circumstances, but we acknowledge his contribution to the colloquium on the *Evolution of Human Wisdom* that took place at Durham University's Institute for Advanced Study in July 2016.
3. Thomas Nagel, *Mind and Cosmos: Why the Materialist Neo-Darwinian Conception of Nature is Almost Certainly False* (New York: Oxford University Press, 2012).
4. Agustín Fuentes, "Human Evolution, Niche Complexity, and the Emergence of a Distinctively Human Imagination," *Time and Mind* 7.3 (2014): 244.
5. Augustine, *De Doctrina Christiana*, ed. and trans. R.P.H. Green (Oxford: Clarendon Press, 1995), II.99.

BIBLIOGRAPHY

Augustine. *De Doctrina Christiana*. Edited and translated by R.P.H. Green. Oxford: Clarendon Press, 1995.

Fuentes, Agustín. "Human Evolution, Niche Complexity, and the Emergence of a Distinctively Human Imagination," *Time and Mind* 7.3 (2014): 241–57.

———. "Integrative Anthropology and the Human Niche: Toward a Contemporary Approach to Human Evolution." *American Anthropologist* 117 (2015): 302–15.
Nagel, Thomas. *Mind and Cosmos: Why the Materialist Neo-Darwinian Conception of Nature is Almost Certainly False.* New York: Oxford University Press, 2012.

Prologue

Philosophical Parameters

ONE

The Human as World-Open Spirit

An Exploration into Philosophical Anthropology and the Foundations of Human Wisdom

Dylan Belton

What follows is conducted primarily in a spirit of experimentation and exploration. I aim to introduce the work of the philosopher Max Scheler into the discussion regarding the evolution of human wisdom. More specifically, I focus on Scheler's notion of the human as world-open spirit (*Geist*). My choice of Scheler will no doubt seem odd not only because of his marginal significance in Anglo-American philosophical circles, but also because of what may be deemed his overly metaphysical and apparently dualistic vision of the human being, as well as his emphasis on a qualitative gap between the human and other animals. Nevertheless, I have chosen to bring Scheler into the discussion for two primary reasons: (1) he is considered a leading figure in the twentieth-century German tradition of *philosophische Anthropologie*, one which I maintain ought to be granted a voice in current discussions on human evolution; and (2) although he died prior to the development of the neo-Darwinian paradigm, Scheler is radically opposed to the vision of the human being as currently presented by certain reductive forms of neo-Darwinism. He is therefore an ally for those seeking to move beyond the latter paradigm.

Before progressing, a few brief comments regarding "wisdom" are in order. I take it for granted that any attempt to explain *what* constitutes human wisdom and *how* it evolved demands a layered approach, involving not only the various life sciences but also disciplines such as philosophy and, ultimately, theology. Traditionally, the term possesses strong normative connotations: one *ought* to pursue wisdom, the wise *ought* to govern, etc. Wisdom also possesses practical and speculative components. That is, the wise individual is not only one who masters practical means towards certain ends but also one who speculates about non-practical matters. The human, for instance, does not simply pursue ends but is the being that speculates *about* ends. For the purposes of this essay, I am interested in what constitutes this speculative component of human wisdom. I leave aside normative questions. While the human exhibits strong continuities with the rest of the animal kingdom, I wish to trouble any attempt to simply reduce human wisdom to an extension of what may be designated as the technical or instrumental rationality that humans share with certain other beings. The human being is uniquely able to perform what Scheler calls "ideation," the act by which "the logos of essences is peeled off from the concrete and sensory world of things."[1] This act is best described as contemplative or speculative, and the capacity to achieve this act demands, according to Scheler's philosophical analysis, the introduction of the notion of spirit as a novel principle.

Although I will focus on the act of ideation, to speak of spirit within Scheler's anthropology is to speak of far more than this act:

> This term [spirit] also contains the concept of "reason," but it encompasses, in addition to the linking of ideas, a special type of intuition of primordial phenomena and essential contents, and it encompasses also a specific class of volitional and emotive *acts* such as kindness, love, repentance, awe, states of wonder, bliss, despair, and free-decision making . . .[2]

It is important to note, however, that I am not utilizing Scheler's philosophical anthropology to negate evolutionary theory or the explanations offered by biologists, anthropologists, and psychologists for the above aspects of the human, but more to suggest that explanations within the sciences require a complimentary metaphysical account, whatever form the latter might take. With that said, I do not have the space in this essay to argue in detail for Scheler's vision: it is simply not possible to do justice here to the development, complexity, and nuance of his thought. I hope, at minimum, that presenting Scheler's view of the human being as world-open spirit will spark questions, protestations, and, hopefully, a sense of wonder at the human species. If I succeed in this, then I consider my task accomplished.

In order to better situate my discussion of Scheler, I outline in the first section a vision of the human being and of human wisdom that emerges within a traditional neo-Darwinian and sociobiological paradigm, which is, in the final analysis, a reductionistic, mechanistic, and adaptationist paradigm shot through with certain philosophical assumptions and convictions. Although many biologists and anthropologists are moving beyond this paradigm, it is important to grasp its underlying logic, which remains a dominant one and which is a logic Scheler's anthropology resists. In the second section, I move on to discuss the vision of the human being as presented by Scheler. By introducing the principle of spirit, Scheler attempts to account for what he calls the human being's *world-openness*. It is this world-openness that ultimately is the human way of being in the world. The evolutionary emergence of the human being's characteristic way of being in the world is thus a process of coming to exhibit this world-openness: "Becoming human is tantamount to being elevated to world-openness by virtue of spirit."[3] By way of conclusion, I will address briefly a few questions that might arise regarding my use of Scheler's notion of spirit.

THE NEO-DARWINIAN PARADIGM

What I am calling neo-Darwinism is the position that accepts what has come to be known as the Modern Evolutionary Synthesis (MS) as developed in the twentieth century, but that also couples this with a commitment to reductionistic and mechanistic philosophical positions regarding not only the nature of organic life but also the nature of mind. Neo-Darwinism is, moreover, essentially an adaptationist program which tends to explicate adaptation primarily (but not exclusively) in terms of individual selection.[4] Sociobiology and evolutionary psychology share its underlying logic. The major achievement of MS was the wedding together of Darwinian natural selection, population-level thinking, and Mendelian inheritance. MS has since gained dominance within scientific circles, leading to the widespread acceptance of several core assumptions about evolution:[5] (1) although genetic drift and gene-flow are important, evolutionarily significant variation stems primarily from genetic mutations; (2) beneficial mutations have small phenotypic effects which result in gradual phenotypic change; (3) inheritance is only genetic; (4) natural selection is the dominant explanation for adaptation; and (5) macro-evolution is the accumulation of differences that arise through micro-evolutionary processes. When certain assumptions regarding the causal role of genes are added to this understanding of evolution, it is not an exaggeration to say that, ultimately, genes and natural selection *constitute* the evolutionary process. Indeed, the evolutionary process was, and still is, described as alterations in allele frequencies within populations over time.[6]

Wedded to a certain understanding of "information," neo-Darwinism maintains that information is somehow "contained" in the genes and encodes for various traits. It is firmly asserted that no new information can enter the genome through Lamarckian-like processes. What this results in is a sharp distinction between the processes of development and heredity. The latter is only a matter of genes while the developmental process itself becomes the "expression" of the information contained *in* the genes. Related to this distinction is a further sharp distinction between "ultimate" (evolutionary) and "proximate" causes.[7] Ultimate evolutionary causes are restricted to natural selection, genetic mutations, gene-flow, and genetic drift. Phenotypic plasticity, non-genetic inheritance, and niche construction, while of serious interest to any biologist and anthropologist, are, within this framework, only proximate causes. In other words, they are considered to be "outcomes" (adaptations) of natural selection and not themselves explanations of organic diversity and adaptation. Again, what *ultimately* matters within this account of evolution takes place at the genetic level: organisms are "programmed" by their genes, which have been shaped by natural selection.[8]

The success of neo-Darwinism was closely tied to movements in the twentieth century toward materialism and reductionism in philosophical circles. Unsurprisingly, the organic system (the teleological system *par excellence*) was "mechanized" insofar as various attempts were made to show how the developmental process itself is, contrary to all appearances, *not* in fact goal directed in any rigorous sense and is, in any case, ultimately irrelevant to evolutionary theory in light of the new genetic theory of heredity.[9] The fact that evolutionarily significant variation is said to stem primarily from *random* genetic mutations played right into the hands of philosophers seeking to rid biology of any teleological principles. The combination of random genetic mutations plus the non-random (but not teleological) principle of natural selection provides evolutionary biology with a non-teleological form of *ultimate* explanation. All of the teleological (adaptive) features of living beings ultimately stem from non-teleological principles (random genetic mutation coupled with natural selection). The distinction between the ultimate and proximate causes is central here, since any scientist immersed in studying actual living beings will be hard pressed not to speak in teleological terms when explaining the activities and the phenotypic make-up of living organisms. Traits are present *because of* the function they serve. Humans, and other organisms, appear as purposive *agents* that act as goal-oriented wholes in shaping their environments. Phenotypic plasticity is also clearly an environmentally sensitive and goal-directed activity of the organism as a whole. By relegating phenotypic plasticity and the niche-constructing activity of organisms to the realm of proximate causes, biologists and philosophers are able to claim that ultimately the evolutionary process is a non-teleological one constituted by the alteration of gene frequencies over time by means of natural selection.

Such a vision has immense consequences for our understanding of organic life in general. Richard Lewontin argues that, within this framework, the organism is now to be conceived of as *passively* molded by the combination of internal (genetic) and external (external environment/natural selection) forces.[10] According to Hans Jonas' analysis of the neo-Darwinian paradigm as it was emerging, mutation at the genetic level represents a mechanical disturbance of the stability of the "gene system." From the viewpoint of the organic system such mutations are mechanical mishaps, and, as such, their phenotypic outcomes are akin to "deformities." This process over time may lead to novel and enriched patterns and organization, as well as to beings more perfectly adapted to their local environments, but such an enrichment will still be an *excresence*: "the high organization of any animal or of man would appear a gigantic monstrosity into which the original amoeba has grown through a long history of disease . . . this must be the picture so long as mutation is conceived as nothing more than a freak accident whose deleterious or beneficial quality is decided only *post hoc* in the lottery of natural selection."[11] One also has to say now that the developed (organism) is for the sake of the undeveloped (genes). While adaptive phenotypes *evolved* over time within a given population because of the beneficial function they serve, the small changes stemming from mutations did not *initially* occur and lead to the beneficial phenotypic outcomes in question within individual organisms *for the sake* of the current function they now serve in relation to the organism as a whole. Natural selection has replaced teleology as a directing principle but does so only by eliminating certain forms thrown up by the "random play of aberrations from pattern . . . on which the distinction between deformity and improvement is superimposed by entirely extraneous criteria."[12]

Jonas also correctly points out how neo-Darwinism represents the final liquidation of immutable essences and signifies the final victory of nominalism over realism, and, by doing so, grants tremendous support to the anti-Platonism of the modern mind: "The minimum left to the original essence of life is just self-preservation, which is analogous to the inertial laws ruling the conduct of a particle."[13] The flux of dynamism has completely replaced the stability of the Aristotelian realm of substantial forms and essences. The classical notion of perfection is therefore rendered an illusion. For Aristotelians, there is such a thing as a definitive pattern of a given species, and there can therefore be a more or less perfect realization of this pattern. The good and wise life is thus one that more perfectly realizes this pattern. Value is built into this ontology. This vision is totally undermined when it is admitted that species are only relatively stable and when each individual organism represents a "trying out" in the drama of adaptation.

Explanations of *human* evolution within the context of neo-Darwinism tend to take on a certain form. Another sharp distinction tends to be drawn between two constitutive "aspects" of the human: biological and cultural.[14] Just as human beings have basal biological (and psychological) traits that are encoded in the genetic material, so do they possess cultural traits that are "encoded" in any given culture prior to the development of a given individual and that subsequently come to be "expressed" within individuals. This position maintains that there is something like a basal biological human nature (including psychological and behavioral traits) on top of which cultural evolution is imposed. This basal human nature has been formed via the ultimate evolutionary processes as described above. And given that these shared species-specific traits must be independent of cultural variations among humans, it is believed that this basal human nature is genetically based. Although currently outdated, it is unsurprising that the notion of the "meme" was developed in order to serve as the equivalent of the gene. Cultural evolution simply becomes an extension of biological evolution but with "memes" rather than "genes" as that which is passed on from generation to generation. When this logic is taken to an extreme, the human is reduced to the assembled product of genetic and cultural instructions.[15]

In order to understand the development of this basal human nature, and hence the basic human cognitive capacities, many seek to understand the adaptive pressures facing the hunter-gatherer communities of the Pleistocene era.[16] The evidence shows that our hominid ancestors faced numerous threats from various predators and what appears to be unusual climate variation that can explain the evolutionary success of increased adaptability.[17] All accounts of the evolutionary emergence of the human being point to an increasing need for, and complexification of, social cooperation in the Pleistocene period due to numerous predator threats, increased hunting and scavenging activity, and cooperative breeding. As is quite well documented, this history is also marked by an increase in hominin brain size, something correlated with increased intelligence and social complexity. Further details regarding the claims made about this evolutionary history need not concern us here. What is important for now is simply the observation that within this framework, what philosophers have traditionally called "reason" is now a part of biology, no different from hands, feet, eyes, etc., which has, as the philosopher Max Horkheimer succinctly puts it, "developed through adaptation to natural conditions and surviving because it [reason] proves to be an adequate means of mastering them, especially in relation to acquiring food and averting danger."[18] The human intellect is utterly "instrumentalized" insofar as its evolution in the human species is due solely to its fitness value, either at the individual or at the group level. It is *essentially* a "tool" for survival and must be explicated as such.

Viewing the human being in this manner fits in perfectly with the discipline of evolutionary psychology, an "adaptationist program applied to the study of the human mind/brain."[19] Evolutionary psychology understands the human mind to be a highly complex input-output mechanism that it divides into various modules, each explained individually in terms of adaptive fitness.[20] Revealing a mode of thinking typical of the framework of neo-Darwinism and of evolutionary psychology, the philosopher of mind David Papineau claims that the various cognitive capacities which comprise the "architecture" of the mind are *accidentally* developed and merely *extrinsically* added on to each other.[21] But such a vision of the human mind raises a host of awkward difficulties. How does one account for the *unity* of these capacities and for the self-conscious "*I*"? What about the human capacity to *overcome* its evolutionarily "programmed" dispositions? As Papineau himself notes, what is unique about the human is its "theoretical rationality" which allows it to *overcome* its evolutionary heritage by "deliberately identify[ing] ways of thinking that are reliable for truth and set[ting] ourselves to practice them."[22] This capacity turns out to be, according to Papineau, nothing but a *spandrel* that "piggy-backs on the evolution of cognitive abilities for 'understanding of mind' and for 'means-end' thinking . . . once these abilities are in place then nothing more is needed for humans to achieve high levels of theoretical rationality."[23] Papineau has to make such a move because what he characterizes as "theoretical rationality" has no "biological advantage" and therefore cannot have been selected for. The human is thus "wise," literally by accident. Importantly, the human is also "wise" in a universe utterly indifferent to value, meaning, or purpose. When E. O. Wilson and Michael Ruse famously claimed that human morality is a mere means for survival and reproduction and hence an illusion if understood to have any more meaning than this, they were stating the logical outcome of the neo-Darwinian logic and its ontology.[24]

THE HUMAN AS WORLD-OPEN SPIRIT

Admittedly, I have not attempted to provide any definitive arguments against the position outlined above. What I am most interested in is its underlying logic. To reiterate, it considers the organic system to be a mechanistic one, and it applies a rigorous adaptationist logic in its explanations. When this logic is extended to the human mind, then the mind is seen as a highly complex input-output mechanism explicable in terms of modules that each serve an adaptive function understood primarily in terms of competition for reproductive success between individuals.[25] One may also rightly point out that the neo-Darwinian paradigm is not as monolithic as I have described above, and hence that there are other more sophisticated ways of being a neo-Darwinian. While this may be the case, it does not change the fact that neo-Darwinism, in the final analysis, will always view the human according to a mechanistic and adaptationist logic such that the human intellect will be considered essentially as a tool for survival. Human wisdom thus will always be explicated as a mere extension of the instrumental intelligence one already finds in certain non-human species.

Although I cannot go into great detail here, it is important to note that we are currently witnessing significant challenges to neo-Darwinism emerging from within biology, as well as from within biological anthropology. A number of biologists and anthropologists are currently pushing for what some are now calling the Extended Evolutionary Synthesis (EES).[26] Central to EES is the focus upon *multiple* layers of inheritance: genetic, epigenetic, behavioral, and symbolic-cultural.[27] Indeed, the very classic notion of the gene as it was developed in the twentieth century has collapsed.[28] The claim that specific genes "program" for specific phenotypes is highly contestable. There is good evidence now as well that some genetic variation is non-random and that epigenetic changes can be passed on to subsequent generations.[29] The gene-centered view of evolution and of the organism is, it seems, giving way to a developmental, process, and systems-centered vision of evolution. The process of evolution is not that of changes in allele frequencies over time but is the emergence of *novel forms of life* within highly relational contexts.[30] Some theorists also seek to place phenotypic plasticity at the forefront of evolutionary theory. West-Eberhard suggests that, rather than random genetic mutations as the key generators of evolutionarily significant variation, this role might actually belong to phenotypic plasticity.[31] According to this account, populations of organisms initially adapt to new environmental pressures through phenotypic plasticity, something that is an entirely non-random process and that might allow for a renewed focus upon Aristotle's notion of hypothetical necessity.[32] Genes subsequently play "catch-up" to solidify the beneficial variations created through plasticity. For some, developments in evolutionary theory challenge the entrenched neo-Darwinian distinction between proximate and ultimate evolutionary causes.[33] Niche-construction is also considered by some to be a central evolutionary process.[34] Organisms are not passively shaped by their environments but actively shape them, thereby playing an active role in generating selection pressures and therefore their own evolutionary trajectory.

These are significant developments and all demand continued philosophical attention. How to incorporate them into a more nuanced understanding of *human* evolution is a project still underway. Suffice it to say that, for now, such developments are opening up a new space for interdisciplinary experimentation and exploration. Moving beyond the neo-Darwinian paradigm is also a philosophical task that requires rethinking the nature of organic life, philosophy of mind, and philosophical anthropology. In the spirit of experimentation and exploration, I turn now to Scheler. Writing just after World War I, Scheler was convinced that only a renewed philosophical anthropology could solve the anxiety of an age of uncertainty regarding the nature and fate of the human species: "In approximately ten centuries of history, this is the first in which man finds himself completely and utterly 'problematical,' in which he no longer knows what he is and simultaneously *knows that* he does not have an answer."[35] Scheler's call for a renewal of philosophical anthropology was due to what he saw as the proliferation of incompatible and incomplete philosophical accounts of the human being. Manfred Frings helpfully summarizes these various other anthropological accounts that were live options at the time:[36] the human as the disease of life (Panromanticism), as the *Superman* (Nietzsche), as *homo sapiens* (Linné), as *L'Homme Machine* (La Mettrie), as the *Power* being (Machiavelli), as the *Libido* being (Freud), as the *Social-Economic* being (Marx), as *Homo Faber* (Bergson, Positivism), and as the *Fallen, Created* being (Christianity). To this list we could add the neo-Darwinian account of the human as described above. Such accounts are, according to Scheler, all too narrow and are not adequate enough to capture the flexibility (*Plastizität*) of the human.[37] Scheler is most opposed to the vision of the human being as *merely* a tool-making animal and therefore consistently resists the *homo faber* account as promoted by both Henri Bergson and the "Positivists" of the day. Scheler labels this latter account the "pragmatic" and "positivistic" account.[38] According to this account, the human is the being who "works and makes tools . . . his 'reason' and 'logic' have framed themselves for such practice."[39] The unique human capacities are nothing but the *extension* of principles operative in the non-human animal realm. Although this Positivist view is not identical to the neo-Darwinian one, the two remain identical in terms of characterizing the emergence of the human being as the mere extension of the "technical intelligence" that is found in other animals.

What exactly is this discipline of philosophical anthropology? It is, as Scheler defines it at one point,

> . . . a basic science which investigates the essence and essential constitution of man, his relationship to the realms of nature (organic, plant, and animal life) as well as to the source of all things, man's metaphysical origin as well as his physical, psychic, and spiritual origins in the world, the forces and powers which move man and which he moves, the fundamental trends and laws of his biological, psychic, cultural, and social evolution, along with their essential capabilities and realities.[40]

This is no doubt a task of astonishing hubris, one which is clearly antithetical to the anti-metaphysical mood of our present age. But it would be a loss to dismiss it on this ground alone. Two more points regarding Scheler's approach to this discipline are important here before moving forward: first, he is operating within the phenomenological tradition which seeks to overcome the reduction of philosophy to epistemology in service of the natural sciences. As one commentator notes, this phenomenological approach is constituted primarily by an attitude of "openness towards" reality "which permits what is revealing itself to reveal itself as it is *in* itself."[41] In *Die Stellung des Menschen im Kosmos*, the text that will be the focus here, Scheler does, however, move beyond mere phenomenology and enters into bold metaphysical and theological speculation.[42] Second, although Scheler speaks of the object of this science as an "essence," this need not be understood in the Aristotelian sense that may raise concern. The aspect of Scheler's thought that I am concerned with has little to do with something like a narrowly normative view of human nature often associated with Aristotelian natural-kind essentialism. But Scheler does believe that he can outline the "essential constitution" of the human being and show how a proper understanding of this constitution is related to metaphysics and theology. This, however, is not to say that he conducts himself in isolation from the natural sciences. Scheler was well versed in the scientific discoveries of his day.

With those introductory remarks in place, I turn now to Scheler's account of the human being. The human as spirit is *world-open* (*weltoffen*), "non-environmental" (*umweltfrei*), no longer subject only to its drives and environment (*Umwelt*).[43] Although certain other animals may be said to possess technical intelligence, they are nevertheless constrained by what is useful or not useful and are also "walled in" by their environment. Animals, in other words, cannot turn the environment into an object. This "walling in" is obviously more intense for the animals that are more instinctually bound. It loosens up when it comes to certain animals that exhibit associative memory and technical intelligence.[44] What happens in the human being, however, is a *detachment* from its organic and psychic "drives." This detachment makes possible the objectification of the environment. Consequently, the human has what Scheler calls "world" (*Welt*). Characteristic of the human as spirit is its "existential detachment from organic being, its freedom and detachability—and the detachment of its center of existence from the bondage to, the pressure of, and the organic dependence on 'life' . . . and thus also its detachment from its own drive-related intelligence."[45] Spirit is "beyond" what is called "life" in its widest meaning. It represents a domain that is irreducible to the inorganic and organic domains of reality. I will take up the question of dualism in the conclusion.

Although Scheler argues that a unique principle (spirit) is required in order to understand the emergence of the human being, this does not mean that he is insensitive to our ontological kinship with other animals. He is extremely generous in his interpretation of other animals and opposes some of the hackneyed ways of contrasting the human to other animals. Animals do not all behave only according to instinct. Some learn through associative memory and can even, through the free combination of ideas, find solutions to unique problems that they have never previously encountered (i.e., they possess practical intelligence). Scheler is thus quite contemporary in terms of his generous interpretation of the capacity of other animals. What separates him from the contemporary milieu is his insistence that, despite the continuities between the human and other animals, a gap is present, and one that, in principle, can never be bridged by any standard Darwinian, physicalist account of human evolution. The gap is what may be called a *metaphysical* one. In other words, the gap can be bridged neither by an anatomical change (e.g., a larger brain) nor by a mere extension of characteristics found in other animals (e.g., an increase in technical intelligence, sociality, etc.) such that human evolution would ultimately be explicable in terms of biology, anthropology, and psychology.

By speaking of the human in terms of spirit and of detachment from organic-bound drives, Scheler is attempting to account for the fact that the human is able, uniquely, to achieve a form of reflecting back upon itself and upon its environment that is not present in other animals. All organic beings possess an ontic center. In animals that possess a nervous system, there is a "reporting or reflecting back" of its organs to a center, i.e., sensation and consciousness. However, in relation to its environment, the animal remains ecstatic in its behavior, and its intelligence remains attached to practical activity (i.e., to its drives). It experiences its own drives as attractions or repulsions emanating from things in its environment. In contrast, the human spiritual act is intrinsically linked to self-consciousness (*Selbstbewußtsein*), to a second level of the act of reflection. Scheler speaks of the human spiritual act as "ingathering" or "concentration" (*Sammlung*) and suggests that this ingathering or concentration, along with its aim of concentration-on-one's-own-self (*Sichsammelns*), be named "the consciousness of the spiritual act-center of itself" or "self-consciousness" (*Selbstbewußtsein*).[46] While this formulation is rather dense, it is clear what Scheler is attempting to convey: in this second act of reflecting back upon itself, the human objectifies not only the external world but its very own physiological and psychic states. Because of this, and because of its spiritual act center, the human is aware of and has a power over itself that other animals do not possess.

The human objectifies insofar as it grasps things in the world *as* such-and-such. This is the *act of ideation,* i.e., the act by which the human distinguishes between essence and existence. It is that act through which *a priori* knowledge is achieved.[47] Scheler is adamant that this act of ideation differs qualitatively from the acts that constitute technical intelligence. Ideation is an *ascetic* act. The human is that being who can, indeed who must, take an ascetic attitude toward its drive impulses: "the human being is the 'Nay-sayer,' he is an ascetic of life.'"[48] To perform this act, there must be a *stepping back from,* or *suspension of,* our normal engagement with the world: to perform this act is to "suspend the element of reality, to annihilate the whole, undivided, and powerful impression of the reality along with its affective correlate."[49] In this ascetic act, the "logos of essences is peeled off from the concrete and sensory world of things—insofar as they have already become 'objects.'"[50] When providing an example of the act of ideation in its most astounding form, Scheler speaks of the Buddha, who, from a few particular instances of suffering, gained insight into the essential makeup of the world. The human can know it is in pain and can investigate pain through the positive sciences. But the human can also step back and ask, from a more "distant, pensive, and contemplative standpoint," the following question: what is "pain (or suffering) itself?" And, as the Buddha did, we are able to ask how pain is related to reality *as a whole*.[51] We may be hopelessly wrong in our answers, but it is the fact that we can achieve this sort of questioning that matters most to Scheler.

As strange as it may sound, Scheler talks of becoming human in terms of a *stepping outside* the entire cosmos, a claim related to his denial that the human is a thing-like-being.[52] Because of this capacity to objectify, the human is characterized by what may be called a capacity for the infinite: "only the human being is able to *soar* far above his status as a living entity and, from a center beyond the spatio-temporal world, make *everything* the object of his knowledge, including himself."[53] The human is thus not at all "walled-in" by the environment. It has a center beyond the spatial-temporal world and therefore appears as unlimitedly open insofar as everything is a potential object of knowledge. The center from which the human acts and which allows it to objectify therefore cannot itself be a "part" of the world of thing-objects. This center, furthermore, cannot itself be rendered an object since it *makes objectification possible*. What this means, according to Scheler, is that spirit's center (sometimes referred to as "person") cannot be thought of as a "thing-like being."[54] As he puts it in a rather dense formulation, "Spirit is the only being which cannot be *objectified*—spirit is *pure actuality*; it exists only in *freely carrying out its acts*. Hence, spirit's center, the 'person,' is not objectifiable, nor is the person a thing-like being. The person is only *in* his acts and exists only *through* them."[55] All of these claims are attempts to clarify his fundamental conviction that what I am calling a certain capacity constitutive of human speculative wisdom cannot be explained merely as an extension of the organic-bound technical intelligence exhibited in other animals.

By introducing this principle of spirit, Scheler is also seeking to ground, at the metaphysical level, what it is that allows the human to pursue disinterested ends and values such as the good, the beautiful, and the just. From such disinterested values "grow great works, comprehensive systems of morality, art and legal systems, all of which place life at their service, and all of which, as elements of civilizations and its technological framework, seek to subjugate life, and consign it to the status of a substructure of their own self-enhancing existence."[56] A non-human animal may demonstrate emotional solidarity with another of its kind, but what it does not have is the "free scope for compassion and spiritual love and kindness *against*, or *independently* of, its own vital interests."[57] While a non-human animal may be able to act spontaneously out of a "center" so as to avoid certain advantages for the sake of others, it cannot choose to prefer certain values over others. Scheler sometimes speaks as if we have a *conflict* between spirit and life, but antagonism is not what he has in mind. Rather, the relation between "life" and "spirit" in the one unified constitution of the human being is that of sublimation: "the human being can sublimate drive-energy into spiritual activity."[58]

CONCLUSION

Far more work of course needs to be done in terms of filling out this vision with reference to evolutionary biology, anthropology, philosophy of organic life and mind, and theology. But, as noted in the introduction, my use of Scheler's anthropological vision has primarily been experimental, exploratory, and evocative. To speak of the evolutionary emergence of the human being within Scheler's framework is to speak of the emergence of world-openness. As such, the human is "never at peace with his environing world; he is always eager to break through the borders of now-here-whatness (*Jetzt-Hier-Sosein*), always [striving] to *transcend* the reality surrounding him, including the reality of his own self."[59] It is the traces of this world-openness that we see deep in the hominin archaeological record. In terms of wisdom, I have tried to focus on one aspect of what I designated speculative wisdom, i.e., the act of ideation, which, according to Scheler's analysis, is no mere extension of the capacities constitutive of the technical intelligence that we see in other animals. This capacity requires that we posit within the human being a "detachment" from its organic bound intelligence. And it is this detachment which, according to Scheler, requires a *metaphysical* (and ultimately theological) explanation along with explanations within the sciences.

I wish to conclude here by highlighting a few pressing questions that will most likely arise in light of what has been covered so far. First, it is crucial to note that Scheler's account as outlined above is by no means a sort of "god-of-the-gaps" argument regarding human origins and capacities. His philosophical anthropology is not an account aimed at *competing* with scientific claims about the human being and human origins. One may claim that such an account is of no use for a scientific understanding of the human being, but this is only relevant if one maintains that science exhausts all necessary "explanation" of human evolution and of human capacities. But this is itself a philosophical claim that requires philosophical defense. Scheler's position, however, categorically does deny that a mere materialist and adaptationist account of the human being can ever fully, in principle, account for the evolution of the human being.

Second, speaking of the human being in terms of existential detachment from organic being, "life," and drive-related, technical intelligence appears to introduce an untenable dualism into anthropology, and it does raise concerns about spirit's embodiment. But Scheler is by no means a dualist in the Platonic or Cartesian sense.[60] As noted, spirit does not by any means designate a "thing-like" being but is rather an unobjectifiable center of continuous acts. If there is any sort of "dualism" here, it is of the Kantian sort, and differentiating between the two is crucial. Nevertheless, Scheler does oppose spirit to nature and life, and it is, he maintains, this opposition that renders philosophy possible. Those who find Scheler's notion of the human as world-open spirit attractive, and who seek to embed evolutionary theory within a non-materialist and non-mechanistic logic and ontology, must therefore seek to incorporate Scheler's insights into a modified framework. One could turn to the work done by later philosophical anthropologists, such as Arnold Gehlen, who sought to render Scheler's anthropological vision less metaphysical.[61] Indeed, along with Jonas, it may even be possible to speak of this world-openness not as an ontological *novum* but as a working out of principles already present in the origins of organic life *per se*.[62] One could also turn to the work of Merleau-Ponty, who, while heavily indebted to Scheler, offers a re-characterization of the relation between life and spirit with an emphasis on the latter's embodiment. Merleau-Ponty's attempt to deal with the "problem of perception" was an attempt to deal with the relation between consciousness as life and consciousness as spirit, and in his later work he came to speak of the relation as one not of sharp division but of an intertwining *Ineinander* (in-one-another).[63] Merleau-Ponty is therefore a crucial resource for those who wish to retain Scheler's insights but who also wish to render his concepts a bit more down to earth.

Finally, in terms of dialogue with theology, Scheler is adamant that the emergence of the human being is simultaneously the emergence of an awareness of the Ground of all being: "it amounts to a total mistake to let Descartes' 'I am' and Saint Thomas Aquinas' 'The world is' precede the proposition of 'There is absolute being' and to deduce the sphere of the absolute from such modes of being. Consciousness of world, of self, and of God form one unbreakable structural unity..."[64] Once the human came to possess the capacity to objectify, to ask about the cosmos as a whole, it could also think of the possibility of absolute nothingness, and this of course prompts the question as to why there is something rather than nothing. "Religious" questioning is thus constitutive of the human being's world-openness. In terms of his account of the relation between finite and Absolute Spirit, he does, in his later *Human Place in the Cosmos*, develop a notion of the *identity* of the human spirit with Absolute Spirit: "For me, the fundamental relation the human being has to the Ground of Being consists in this Ground grasping itself and realizing itself in the human being who is, because he has both spirit and life, a part of the Ground's own spirit and impulsion."[65] Theologians who believe that this notion of spirit still warrants attention would have to address these theological moves.

NOTES

1. Max Scheler, *The Human Place in the Cosmos*, trans. Manfred S. Frings (Evanston, IL: Northwestern University Press, 2009), 37.
2. Ibid., 26.
3. Ibid., 28.
4. According to Agustín Fuentes, "many neo-Darwinian approaches assume that the individual and its relationship to the environment (assessed ultimately via genetic fitness tradeoffs) is the most salient feature of a system. Even when there is explicit acknowledgment of a key role for social groups and institutions, the basal assumption is that evolutionarily relevant processes ultimately result from the competition for reproductive success between individuals (as mediated by their environment and social groups). This creates a theoretical context wherein the individual and its fitness-based relationship to a given (social/ecological) environment is the basal heuristic framework for evolutionary models." Agustín Fuentes, "The Extended Evolutionary Synthesis, Ethnography, and the Human Niche: Toward an Integrated Anthropology," *Current Anthropology* 57.S13 (2016): 14.
5. Kevin N. Laland et al. "The Extended Evolutionary Synthesis: Its Structure, Assumptions, and Predictions," *Proceedings of the Royal Society B* 282.1813 (2015): 2.
6. To take but one instance of this view, E. O. Wilson writes: "The fundamental evolutionary event is a change in the frequency of genes and chromosome configurations in a population." E. O. Wilson, *The Diversity of Life* (New York: W. W. Norton & Company, Inc., 1992), 75.
7. See Ernst Mayr, "Cause and Effect in Biology," *Science* 134.3489 (1961): 1501–6.
8. Ibid., 1502.
9. See Ron Amundson, *The Changing Role of the Embryo in Evolutionary Thought* (New York: Cambridge University Press, 2007), 159–69. The literature regarding teleology is immense, and the discussion is still a live one, especially since the publication of Thomas Nagel's recent book *Mind and Cosmos: Why the Materialist Neo-Darwinian Conception of Nature is Almost Certainly False* (New York: Oxford University Press, 2012). For a classic treatment of teleology within the context of the modern synthesis, see Ernst Mayr, *Toward a New Philosophy of Biology* (Cambridge, MA: Belknap Press of Harvard University Press, 1988). The burgeoning field of cybernetics in the mid-twentieth century also helped philosophers in their attempt to reduce what appears to be genuine goal-directedness in the biological realm to purely mechanistic processes. See Mark Bedau, "Where's the Good in Teleology?" *Philosophy and Phenomenological Research* 52:4 (1992): 781–806.
10. Richard C. Lewontin, "Gene, Organism, and Environment," in *Cycles of Contingency*, ed. Susan Oyama et al. (Cambridge, MA: MIT Press, 2001), 59–66.
11. Hans Jonas, "Philosophical Aspects of Darwinism," *The Phenomenon of Life* (Evanston, IL: Northwestern University Press, 2001), 51.
12. Ibid.
13. Ibid., 46.
14. Tim Ingold, "Against Human Nature," in *Evolutionary Epistemology, Language, and Culture*, ed. N. Gontier, J.P. van Bendegem, and D. Aerts (Dordrecht, Netherlands: Springer, 2010), 259–81.
15. This view, according to Ingold, ultimately results in the vision of human beings as "trait-bearing cultural clones whose only role in life is to express—in their behavior, artefacts, and organizations—information that has been transmitted to them from previous generations." Tim Ingold, "The Trouble with 'Evolutionary Biology,'" *Anthropology Today* 23.2 (2007): 14.
16. For a good recent explication of this approach to human evolution, see Carol P. van Schaik, *The Primate Origins of Human Nature* (Hoboken, NJ: John Wiley & Sons, Inc., 2016).
17. See Susan C. Anton and J. Josh Snodgrass, "Origins and Evolution of Genus *Homo*: New Perspectives," *Current Anthropology* 53.S6 (2012): S479–S496.

18. Max Horkheimer, *The Eclipse of Reason* (New York: Oxford University Press, 1947), 85.

19. See Kevin N. Laland and Gillian R. Brown, *Sense and Nonsense: Evolutionary Perspectives on Human Behavior*, 2nd ed. (Oxford: Oxford University Press, 2011).

20. Evolutionary psychologists also believe firmly in what they call "the psychic unity of mankind," which can be explained within an adaptationist framework. See John Tooby and Leda Cosmides, "The Psychological Foundations of Culture," in *The Adapted Mind: Evolutionary Psychology and the Generation of Culture*, ed. Jerome H. Barkow, Leda Cosmides, and John Tooby (New York: Oxford University Press, 1992), 19–136.

21. David Papineau, *The Roots of Reason: Philosophical Essays on Rationality, Evolution, and Probability* (Oxford: Oxford University Press, 2006). Passages such as these are therefore common: "The means-end system is simply added *on to the side* of the existing stimulus-response architecture, as it were, leaving the rest as before" (ibid., 68, emphasis added); ". . . means-end reasoning is an evolutionary *add-on*, an extra way of adjusting stimulus-response links" (ibid., 69, emphasis added).

22. Ibid., 60.

23. Ibid., 40.

24. Michael Ruse and Edward O. Wilson, "Moral Philosophy as Applied Science," *Philosophy* 61.236 (1986): 173–92.

25. As Agustin Fuentes notes, even when neo-Darwinism nuances its position by acknowledging the role of social groups and institutions in human evolution, its root assumption still remains that "evolutionary relevant processes ultimately result from the competition for reproductive success between individuals" ("The Extended Evolutionary Synthesis," 14).

26. See Agustin Fuentes, "Blurring the Biological and Social in Human Becomings," in *Biosocial Becomings: Integrating Social and Biological Anthropology*, ed. Tim Ingold and Gisli Palsson (New York: Cambridge University Press, 2013), 59–83; and "The Extended Evolutionary Synthesis."

27. Eva Jablonka and Marion J. Lamb, *Evolution in Four Dimensions* (Cambridge, MA: MIT Press, 2005).

28. See the work of Evelyn Fox Keller, especially *The Century of the Gene* (Harvard: Harvard University Press, 2002) and *Making Sense of Life* (Harvard: Harvard University Press, 2002).

29. See Eva Jablonka and Gal Raz, "Transgenerational Epigenetic Inheritance: Prevalence, Mechanisms, and Implications for the Study of Heredity and Evolution," *Quarterly Review of Biology* 84.2 (2009): 131–76.

30. Anthropologist Tim Ingold has been pushing for this sort of revision for years now: "Forms of life, then, are neither genetically nor culturally preconfigured but emerge as properties of dynamic self-organization of developmental systems. And evolution is their derivational history." Tim Ingold, "Prospect," in *Biosocial Becomings*, ed. Ingold and Palsson, 9, emphasis added.

31. Mary J. West-Eberhard, *Developmental Plasticity and Evolution* (New York: Oxford University Press, 2003).

32. See Denis Walsh, "Teleology," in *The Oxford Handbook of Philosophy of Biology*, ed. Michael Ruse (New York: Oxford University Press, 2008): 113–33.

33. See Kevin N. Laland et al., "Cause and Effect in Biology Revisited: Is Mayr's Proximate-Ultimate Dichotomy Still Useful?" *Science* 334.6062 (2011): 1512–16.

34. See Emma Flynn et al., "Developmental Niche Construction," *Developmental Science* 16.2 (2013): 296–313.

35. Max Scheler, "Man and History," in *Philosophical Perspectives*, trans. Oscar A. Haac (Boston: Beacon Press, 1958), 65.

36. Manfred Frings, *Max Scheler*, 2nd ed. (Milwaukee: Marquette University Press, 1996), 22.

37. Scheler is well aware of human diversity, and so avoids any parochial definition of the human: "Man is a thing so broad, variegated and diverse that all definitions

turn out a little too concise. He has too many ends!" Max Scheler, "On the Idea of Man," trans. Clyde Nabe, *Journal of the British Society for Phenomenology* 9.3 (1978): 186.

38. Scheler, "Man and History," 76.

39. Scheler, "On the Idea of Man," 186.

40. Scheler, "Man and History," 65.

41. Arthur Luther, "The Articulated Unity of Being," in *Max Scheler (1874–1928): Centennial Essays*, ed. Manfred S. Frings (The Hague: Martinus Nijhof, 1974), 4.

42. For a critical account of the relation between Scheler's phenomenological method and his metaphysical anthropology, see Eugene Kelly, *Structure and Diversity: Studies in the Phenomenological Philosophy of Max Scheler* (Dordrecht: Kluwer Academic Publishers, 1997).

43. Scheler, *Human Place*, 27.

44. He is emphatic that an animal is "neither a drive-mechanism nor one of instincts, associations, and reflexes." Ibid., 25.

45. Ibid., 27.

46. Ibid., 29.

47. In one definition, ideation refers to the "grasp of the *essential* qualities and structures of the world in terms of *one* example of some essential region, independently of the magnitude and number of observations and inductive reasoning from them." Ibid., 36.

48. Ibid., 39. As he puts it in this same section, "The human being is a creature that, by virtue of its spirit, can take an *ascetic* attitude toward its fervent and vibrating life."

49. Ibid.

50. Ibid., 37. Although this stepping back is equivalent to Husserl's *epoché*, Scheler did not quite follow Husserl in his understanding of the phenomenological reduction. However, he agrees with Husserl that this act defines the human spirit (ibid., 38–39).

51. Ibid., 35.

52. Ibid., 63.

53. Ibid., 33.

54. Manfred Frings explains this well: "Since all things, including man himself, are raised to objectification, spirit itself *cannot* be part of this objectified world, nor can it be an object in itself (in the sense of thing-object), because spiritual being is the *act* of *making* a thing into an object, an act of raising world (in space and time) to objectification. Spirit, then, cannot have its foundation or source in this objective world . . . " Frings, *Max Scheler*, 17.

55. Scheler, *Human Place*, 34.

56. Max Scheler, *The Constitution of the Human Being*, trans. John Cutting (Milwaukee: Marquette University Press, 2008), 140.

57. Ibid., 141.

58. Scheler, *Human Place*, 40.

59. Ibid., 40.

60. He in fact considers spirit to be "impotent." It is thus by no means a Cartesian-like soul that injects energy or motion into the universe. This is a controversial aspect of Scheler's philosophy, and interpretations are hampered by the fact that he died before he could develop his account of spirit.

61. Arnold Gehlen, *Man: His Nature and Place in the World*, trans. Clare McMillan and Karl Pillemer (New York: Columbia University Press, 1988).

62. See, especially, Jonas, *Phenomenon*.

63. On the relation between Scheler and Merleau-Ponty, see Ted Toadvine, "'Strange Kinship': Merleau-Ponty on the Human-Animal Relation," *Analecta Husserliana* 93 (2007): 17–32.

64. Scheler, *Human Place*, 63.

65. Ibid., 65.

BIBLIOGRAPHY

Amundson, Ron. *The Changing Role of the Embryo in Evolutionary Thought*. New York: Cambridge University Press, 2007.

Anton, Susan C., and Josh Snodgrass. "Origins and Evolution of Genus *Homo*: New Perspectives." *Current Anthropology* 53.S6 (2012): S479–S496.

Bedau, Mark. "Where's the Good in Teleology?" *Philosophy and Phenomenological Research* 52.4 (1992): 781–806.

Brown, Gillian R., and Kevin N. Laland. *Sense and Nonsense: Evolutionary Perspectives on Human Behavior*, 2nd edition. Oxford: Oxford University Press, 2011.

Cosmides, Leda, and John Tooby. "The Psychological Foundations of Culture." In *The Adapted Mind: Evolutionary Psychology and the Generation of Culture*, edited by Jerome H. Barkow, Leda Cosmides, and John Tooby, 19–136. New York: Oxford University Press, 1992.

Flynn, Emma G., Kevin N. Laland, Rachel L. Kendal and Jeremy R. Kendal. "Target Article with Commentaries: Developmental Niche Construction." *Developmental Science* 16.2 (2013): 296–313.

Frings, Manfred S. *Max Scheler*, 2nd edition. Milwaukee: Marquette University Press, 1996.

Fuentes, Agustín. "Blurring the Biological and Social in Human Becomings." In *Biosocial Becomings: Integrating Social and Biological Anthropology*, edited by Tim Ingold and Gisli Palsson, 59–83. New York: Cambridge University Press, 2013.

———. "The Extended Evolutionary Synthesis, Ethnography, and the Human Niche: Toward an Integrated Anthropology." *Current Anthropology* 57.S13 (2016): S13–S26.

Gehlen, Arnold. *Man: His Nature and Place in the World*. Translated by Clare McMillan and Karl Pillemer. New York: Columbia University Press, 1988.

Horkheimer, Max. *The Eclipse of Reason*. New York: Oxford University Press, 1947.

Ingold, Tim. "Against Human Nature." In *Evolutionary Epistemology, Language, and Culture*, edited by Nathalie Gontier, Jean Paul Van Bendegem, and Diederik Aerts, 259–81. Dordrecht: Springer, 2010.

———. "Prospect." In *Biosocial Becomings: Integrating Social and Biological Anthropology*, edited by Tim Ingold and Gisli Palsson, 1–21. New York: Cambridge University Press, 2013.

———. "The Trouble with 'Evolutionary Biology.'" *Anthropology Today* 23.2 (2007): 13–17.

Jablonka, Eva, and Gal Raz. "Transgenerational Epigenetic Inheritance: Prevalence, Mechanisms, and Implications for the Study of Heredity and Evolution." *Quarterly Review of Biology* 84.2 (2009): 131–76.

Jablonka, Eva, and Marion J. Lamb. *Evolution in Four Dimensions*. Cambridge, MA: MIT Press, 2005.

Jonas, Hans. *The Phenomenon of Life*. Evanston, IL: Northwestern University Press, 2001.

Keller, Evelyn Fox. *The Century of the Gene*. Harvard, MA: Harvard University Press, 2002.

———. *Making Sense of Life*. Harvard, MA: Harvard University Press, 2002.

Kelly, Eugene. *Structure and Diversity: Studies in the Phenomenological Philosophy of Max Scheler*. Dordrecht: Kluwer Academic Publishers, 1997.

Laland, Kevin N., Kim Sterelny, John Odling-Smee, William Hoppitt, and Tobias Uller. "Cause and Effect in Biology Revisited: Is Mayr's Proximate-Ultimate Dichotomy Still Useful?" *Science* 334.6062 (2011): 1512–16.

Laland, Kevin N., Tobias Uller, Marcus W. Feldman et al. "The Extended Evolutionary Synthesis: Its Structure, Assumptions, and Predictions." *Proceedings of the Royal Society B* 282.1813 (2015): 1–14.

Lewontin, Richard C. "Gene, Organism, and Environment." In *Cycles of Contingency*, edited by Susan Oyama, Paul E. Griffiths, and Russel D. Gray, 59–66. Cambridge, MA: MIT Press, 2001.

Luther, Arthur. "The Articulated Unity of Being." In *Max Scheler (1874–1928): Centennial Essays*, edited by Manfred S. Frings. The Hague: Martinus Nijhof, 1974.

Mayr, Ernst. "Cause and Effect in Biology." *Science* 134.3489 (1961): 1501–6.

———. *Toward a New Philosophy of Biology*. Cambridge, MA: Belknap Press of Harvard University Press, 1988.

Nagel, Thomas. *Mind and Cosmos: Why the Materialist Neo-Darwinian Conception of Nature is Almost Certainly False*. New York: Oxford University Press, 2012.

Papineau, David. *The Roots of Reason*. Oxford: Oxford University Press, 2006.

Pinker, Steven. *How the Mind Works*. New York: W.W. Norton & Company, 1997.

———. *The Language Instinct*. New York: Harper Perennial Modern Classics, 1994.

Ruse, Michael, and Edward O. Wilson. "Moral Philosophy as Applied Science." *Philosophy* 61.236 (1986): 173–92.

Scheler, Max. *The Constitution of the Human Being*. Translated by John Cutting. Milwaukee: Marquette University Press, 2008.

———. *The Human Place in the Cosmos*. Translated by Manfred S. Frings. Evanston, IL: Northwestern University Press, 2009.

———. "Man and History." In *Philosophical Perspectives*. Translated by Oscar A. Haac. Beacon Hill, Boston: Beacon Press, 1958.

———. "On the Idea of Man." Translated by Clyde Nabe. *Journal of the British Society for Phenomenology* 9.3 (1978): 184–98.

Toadvine, Ted. "'Strange Kinship': Merleau-Ponty on the Human-Animal Relation." *Analecta Husserliania* 93 (2007): 17–32.

Van Schaik, Carol P. *The Primate Origins of Human Nature*. Hoboken, NJ: John Wiley & Sons, Inc., 2016.

Walsh, Denis. "Teleology." In *The Oxford Handbook of Philosophy of Biology*, edited by Michael Ruse, 113–33. New York: Oxford University Press, 2008.

West-Eberhard, Mary J. *Developmental Plasticity and Evolution*. New York: Oxford University Press, 2003.

Wilson, Edward O. *The Diversity of Life*. New York: W.W. Norton & Company, Inc., 1992.

I

Signs in Evolutionary Anthropology

TWO

What Can Anthropology Say about the Evolution of Human Wisdom?

Marc Kissel

While on the way to the Evolution of Human Wisdom colloquium, I received a phone call from an anthropologist. When I told her where I was going, her reply was, "Wisdom? Is that something that anthropologists even study?" This is not an unexpected response. Anthropologists, especially those on the biological and archaeological side of anthropology, do not often conceptualize human behavior in terms of wisdom. While some of our terms are often complex and heavily nuanced (i.e., "the ontological turn"), there is much push back against this word from my colleagues, perhaps due to it seemingly being value-laden.[1]

On the other hand, our species name, *Homo sapiens*, reflects the deep-seated belief that wisdom is in fact what makes us human. We refer to an earlier member of our genus as *Homo habilis*, which suggests that it was the species' skillfulness that allowed the species to succeed, and that *Homo erectus*' upright posture gave it a significant advantage. Linneaus, when creating a place for humans in his naming system, defined his own species not based on morphological characteristics but rather by the enigmatic Latin "nosce te ipsum" ("know thyself").[2] In later editions, he notes that it is reason which separates humans from the other primates. Thus, we implicitly see wisdom as *the* characteristic that made us human, even if as anthropologists we are reticent to use that phrase.

There is also a benefit in striking out into new territory. Recent publications have challenged much of what anthropologists in the past have considered to be the major transitions in evolutionary history.[3] It is clear that, for example, many of what we see as the critical events in the past, such as the Middle-to-Upper Paleolithic revolution, are more a product of convention, history, and research agendas than behavioral differences between groups.[4]

Nevertheless, since the 1980s there has been a concern that even though our species appeared ~200,000 years ago, they were not fully behaviorally modern until well after this time.[5] This so-called Sapiens-Paradox[6] has produced copious amounts of research and debate.[7] Here, I want to point out four implications of this hypothesis: 1) this debate assumes that there was a time when we looked fully human, but did not act human; 2) it suggests that there is going to be a clear, and archaeologically visible, dividing line between humans who act in a modern way and those who do not; 3) that the cause of this change was genetic;[8] and 4) if it was genetic this means that there was a fitness advantage to this new behavior.

In this chapter, I explore how anthropologists can use our datasets (genetic, archaeological, cultural, and fossil) to interrogate the question of how our species became behaviorally modern. Concentrating on the archaeological and fossil evidence, I suggest that we have reached an interesting place in that the barriers to understanding the humanization of hominins comes from how we approach the question in the first place. Rather than looking for a time when we became symbol-users, we need to think more about how early human populations were acting, and reacting, in response to internal and external pressures. Becoming human was a long-term process, which involved changes in our phenotype, genotype, behavioral repertoire, and sociocultural niche.

With this in mind, I want to sketch out some ways in which we can use the anthropological record to look for when humans began to act in ways that we would see as "human-like." To begin with, we must note that the phrase "modern behavior" is in itself problematic.[9] As our discussions of the archaeological record will make clear, discerning when we began to act in a "modern"[10] way is very difficult. However, I will argue that there is a way forward though the uses of semiosis. But first we must investigate what other sorts of data are available.

GENETICS

Are there genes for modern human behavior?[11] Writing in 2000, archaeologist Richard Klein commented:

> When the full sweep of human evolution is considered, it is surely reasonable to propose that the shift to a fully modern behavioral mode and the geographic expansion of modern humans were also coproducts of a selectively advantageous genetic mutation. Arguably, this was the most significant mutation in the human evolutionary series, for it produced an organism that could alter its behavior radically without any

> change in its anatomy and that could cumulate and transmit the alterations at a speed that anatomical innovation could never match. As a result, the archeological record changed more in a few millennia after 40 ky ago than it had in the prior million years, and the long-standing pattern of coevolution between anatomy and behavior was ruptured.[12]

Klein and others often point to the FOXP2 gene[13] as an example of a mutant gene that could produce major changes in phenotype. This gene is found on chromosome 7 in many mammals and is highly conserved, with the human version of the gene differing from that of the mouse by only three amino acids.[14] What makes the gene of interest to anthropologists is a mutation found in a small family in London. This unique mutation seems to be the cause of a language disorder that produces difficulties with grammar and vocalizations. It has thus been considered to be a type of "language gene" and is often considered to be a "critical gene known to underlie speech."[15]

As the same version of the gene has been found in Neandertals and Denisovans, the "human" version has deep roots.[16] It might not be surprising that this gene is old. But it does not necessarily tell us about when language first appeared. Interestingly, the region around the gene shows low variability, which is a sign that there has been selection on that region over the last 100 ka.[17] Recent work suggests that *most*, but not all, modern humans have a mutation in the region surrounding FOXP2 and this gene change may influence speech or other aspects of brain function.[18] But since there are living people today who share the ancestral version it is unlikely that this gene was the *cause* of language.

The sequencing and study of ancient DNA has revolutionized the field of paleoanthropology to the extent that it is almost impossible to keep up with new results. What we can say is that it is likely that there were multiple hominin populations leading up to, and perhaps following, the traditional time periods that bookend anatomical and behavioral modernity. But for Klein's model to be correct we would have to find a gene that is found in all modern humans and not in any of the pre-50 kya samples. Evolution has sped up over the last 10,000 years due in part to changes in demography.[19] This would suggest that we are never going to be able to pinpoint a specific gene that allowed hominins to become human.

THE FOSSIL RECORD

It is difficult to pinpoint when *Homo sapiens* first arose. Fossil evidence suggests that by 195 ± 5 kya, fossils in East Africa fit into the species *Homo sapiens*.[20] The Herto material, consisting of a child and two adults, and dating to 160–154 kya, suggests that modern human anatomy was not yet fully evolved, as the variation seen in these fossils is not quite within the range seen in modern humans, at least according to its describers.[21] Because of this, some prefer the subspecies name *Homo sapiens idaltu*, as a paleospecies intermediate between earlier pre-*sapiens* and later *H. sapiens sapiens*. While it is not clear how useful this designation is, recent overviews suggest to some that *Homo sapiens*, strictly defined, are no more recent than 200 kya and that even some post-200 kya fossils may not fit a strict definition.[22] Others, however, argue that *H. sapiens* evolved significantly earlier than 200 kya.[23]

Often paleoanthropologists speak of Anatomically Modern Humans (AMH). As its exact definition is unclear, determining which fossils to place into the AMHs has proved to be quite complex, especially as many of them come from uncertain contexts. Writing in 1995, P. V. Tobias noted a problem that has largely gone unrecognized:

> Instead of speaking of "anatomically modern human beings," we should speak of "modern" or "recent" humans, using a temporal not a morphological descriptor . . . After all, we are not in the habit of speaking of "anatomically modern elephants" or "anatomically modem hippopotami!"[24]

In other words, concepts like AMH, while useful, are also hard to define in practice. The search for shared-derived traits that fit only *H. sapiens sapiens* has produced much interest, but we are still hard-pressed to discern what exactly makes a fossil anatomically modern.

Thus, in order to properly understand the "process" of modern human origins,[25] it is necessary to understand the demographic nature of the Middle Paleolithic/Middle Stone Age. This "muddle in the middle"[26] has become only more confusing over time. The fossil sample from this time period has yielded various hypotheses regarding the speciocity of the Middle Pleistocene. Numerous hominin species have been proposed to account for the large number of fossil finds and the variation within this sample. While most scholars have concentrated their attention on *Homo heidelbergensis* and *Homo erectus*, other plausible candidates include *Homo rhodesiensis*, *Homo mauritanicus*, and *Homo helmei*.

If there are indeed multiple hominin species during this time period, perhaps this radiation can explain the origins of behavioral modernity. Properly contextualizing the Middle Pleistocene record, both skeletally and behaviorally, is necessary to understand what happened next. Yet, there is little correlation between biological anthropology and archaeology (let alone sociocultural anthropology, which is often overlooked). The relevance of an anthropological perspective in understanding this process is critical.

In a series of publications, Marta Lahr and Richard Foley[27] argue that *H. helmei* evolved 300 kya in Africa and dispersed from Africa into Europe around 250 kya, and it is *this* group of hominins that deserve the distinction of being the last common ancestor of both Neandertals and modern humans. While provocative, exactly how this new species is distinguished morphologically is unclear. *H. helmei* lacks a formal morphological description. Possible candidates for inclusion in the *H. helmei* hypodigm include the Florisbad, Jebel Irhoud, and Eliye Springs fossils from Africa, and the Sima de los Huesos, Steinheim, and Swanscombe samples from Europe.

There are a number of factors which suggest that the *H. helmei* species is not valid on a morphological basis. For one, it appears in the record *after* Neandertal characteristics begin to appear in Europe. The fossils that make up the proposed species are also highly variable in morphology. The Eliye Springs cranium, for example, is pathological, making comparative morphological assessments difficult.[28] The significant question is, what does this variability in the fossils mean? Does it suggest a single species, or is it rather evidence of regional differentiation under a multiregional model? Further, some of these crania are now known to date to a period before the apparent introduction of *H. helmei.* One solution would be to remove these from the hypodigm,[29] while another would be to extend its presence further into the past. The dating of the Sima de los Huesos fossils suggests a range from 500–400 kya and the fossils are found in association with one hand-axe (nicknamed "Excalibur").[30] Some scholars believe that the hominins were intentionally placed in the pit and that the hand-axe, which is made of a rare stone, indicates a form of symbolic behavior. While the date is close to Foley and Lahr's suggested age, it does not fit their model in a convincing manner.

The framework for Foley and Lahr's hypothesis is Grahame Clark's stone tool typologies.[31] He posited that archaeologists could study the evolution of stone tool technologies in the same way as biologists study the evolution of organisms. Echoing the calls of Processual Archaeology, Clark reassured archaeologists that they could use techniques similar to ones used by biologists. It is not clear exactly how analogous Clark thought these two processes are, but later archaeologists have suggested that stone tools are subject to drift, flow, natural selection, and mutation.[32]

The exact correlation between the methods of biological science and archaeology, though, should be seriously scrutinized. There is no reason to assume that biological concepts should work better than concepts from the social sciences, such as agency or practice theory, when studying material culture. Biological forces are important, of course, but they must be considered together with concepts derived from ethnography, history, sociology, and psychology, all of which are neglected in Clark's model. If archaeologists want to study stone tools through biological evolution, then it needs to be clear what exactly the method of transmission could be. Stone tools, after all, do not breed. Clark's typology does not include regional differences, and wide-ranging typologies such as Clark's tend to miss the variation by their focus on uniformity, as specific behavior at particular sites and regions is overlooked. We cannot forget that advanced hominins could respond to stresses and situations in ways that other animals cannot. Further, his typology does not take into account raw material availability, which can affect the evolution and specific *chaîne opératoire* of a technology.[33]

The important point for this discussion is that just because objects can be sorted into groups, this does not mean that they represent actual conceptual divisions. A problem arises when archaeologists forget that they are the ones who created these divisions in the first place and fail to ask if these categories represent an archaeological reality.

While there are problems with Foley and Lahr's proposed phylogeny, it is also an incredibly useful one: it helps to explain why Neandertals are so smart. The first appearance date of *H. helmei* corresponds roughly to when the archaeological record begins to show more definitive evidence of symbolic thought.[34] While not originally presented in this way, this correspondence heightens the relevance of the hypothesis, as it not only explains the florescence of modern human culture but also may explain the recent evidence of Neandertal symbolic thought at well.[35] While it may not fit all the available data, it does, in a way other models do not, explain why both populations are creating these artifacts.

ARCHAEOLOGY

Archaeologists in the early part of the twentieth century often created trait lists, aimed at determining how to situate specific archaeological cultures within the neo-evolutionary framework.[36] As part of the processual critique, the trait list approach was argued to be invalid. Archaeologists quickly began to argue against such lists.[37] Defining what makes an archaeological state, for instance, is less a matter of checking boxes and more the recognition of how processes over time lead to greater complexity and then how this complexity manifests itself.[38] While calls for the rejection of a trait list approach are far from uncommon, there still exists the implicit notion that the origins of modern human behavior were a series of events rather than a process. For scholars of human evolution it is still common to see clear separations/disjunctions in the evolutionary record rather than gradients, at least in terms of behavior.

One reason for this may be the influence of Leslie White (1900–1975), whose theories on cultural evolution influenced a generation of anthropologists. Writing in 1940, White argued that the symbol was "the basic unit of all human behavior and civilization."[39] White suggested that the difference between humans and the apes is not one of degree but one of kind:

> There is a fundamental difference between the mind of man and the mind of non-man. This difference is one of kind, not one of degree.

> And the gap between the two types is of the greatest importance—at least to the science of comparative behavior. Man uses symbols; no other creature does. A creature either uses symbols or he does not; there are no intermediate stages.[40]

Much of the discussion seen today, and the debates therein, is derived from this framework. White is disagreeing with Darwin and others who argued that evolution, and especially cultural evolution, shows that there are differences between humans and the rest of the species by degrees. White is arguing for a type of human exceptionalism that would be rejected by most modern anthropologists. However, paleoanthropologists still often assume that symbolic thought is a "you have it or you do not" trait. It thus suggests that there is a major gap between modern humans and earlier, non-symbolic species, an "us versus them" mentality that sees these differences as facts of nature rather than aspects of historical and cultural processes.[41]

For some scholars, the best currently accepted date for what has been called modern human behavior is after 100 kya.[42] Recent research has centered on ochre use,[43] perforated marine shells,[44] and the creation of a complex tool kit, including the origins of bone[45] and composite[46] tools. Thus, it is argued that modern humans, with behaviorally modern traits, did not evolve until well after *H. sapiens* evolved, leaving a time gap between these two events. Explaining the paradox has been a central question in evolutionary anthropology since the mid-1980s with the Human Revolution conference, which helped to popularize the notion of symbolic thought as the distinctive trait of modern humans over Neandertals. One of the first archaeologists to produce a trait list of what makes us human compared to previous hominins was Paul Mellars.[47] He argued for major differences between the Middle and Upper Paleolithic in terms of the archaeological record, which was used to suggest significant behavioral differences. Sally McBrearty and Alison Brooks[48] warned paleoanthropologists about accepting a "human revolution" model, suggesting that the package of behavioral modernity did not arise suddenly.

Chris Henshilwood and Curtis Marean suggest the term "fully symbolic *sapiens* behavior," as it emphasizes the uniqueness of *H. sapiens* compared to Neandertals.[49] Even so, in some sense the problem is that their term *assumes* that there must be something unique about AMH behaviorally that sets them apart. As Trenton Holliday[50] notes, the displacement of coyotes by wolves at Yellowstone National Park is not seen as being a result of wolves' greater intellectual abilities,[51] but this is assumed to be the case at times when modern humans apparently outcompeted other human populations. Furthermore, by assuming that there are important behavioral differences, it is not difficult to find behaviors that Neandertals did not engage in, and that Middle Stone Age (MSA, ~300,000 to ~50,000, depending on location) and Later Stone Age (LSA,

~50,000 to ~10,000, depending on location) humans did, that become the defining feature of modern/fully cultural/symbolically organized behavior.

For scientists who differentiate between Neandertals and modern humans with the designations *Homo neanderthalensis* and *Homo sapiens*, the Latin names seemingly signify a difference in intelligence. However, it would be interesting to ask if one can uncover derived behaviors—that is, behaviors that are not primitive but associated with the evolution/appearance of a new species or type—in Neandertal sites that are not recorded in modern human sites. If that were the case, what would such behaviors suggest? As Neandertals are often used as an example of the "other,"[52] a group which is not quite human and thus used in juxtaposition to show what real human behavior is, identifying such behaviors could help us to better understand what is meant by modern human behavior.

Just exactly which fossils fit into the Neandertal group is a complex one. A conservative approach would classify as belonging to this category human fossils and sites from a period between 130,000 and 30,000 years ago, the time of the classic, or Würmian, Neandertals. The complexities of the Middle-to-Late Pleistocene remains in Europe may indicate that they represent more than one species. In this case, the fossils from this time period do not demonstrate genetic continuity but rather multiple migration events from different regions in small, isolated populations.[53] Yet, demonstrating gene flow would be difficult for such a small sample size (the number of fossils for this time period number being less than twenty, not counting the twenty-seven from Sima de Los Huesos). The alternative hypothesis, that they represent a random breeding population, has yet to be falsified.

The question of who is a Neandertal often rests on the specific definition of the group. While the majority of scholars still conceive of them as separate populations (i.e., "*Homo neanderthalensis*"), both fossil and genetic data suggest a more nuanced approach. Due to the recognition of significant gene flow between the Vindija Neandertals and *H. sapiens*,[54] the null hypothesis of one species may be more defensible, but it leaves open the question of when the Neandertal population first began to diversify. It is clear that neither a strict climatic nor a cladistics approach can capture the complex picture of the Neandertalization process.

While originally considered either to be unable to create artwork or simply to be copying modern humans without really understanding what they were doing, Neandertals are now known to have mixed together pigments and used marine shells as palates.[55] Recent U-series dating of cave art from Spain has supported the assertion that Neandertals may have engaged in artwork. Several large red disks, painted by blowing, have been found in the *Panel de las Manos* in the El Castillo cave.[56]

To investigate the question of Neandertal behavior, I examined reports of the non-dietary use of birds at Middle Paleolithic Neandertal sites and at Middle Stone Age sites associated with AMH. Later Neandertals at various sites have been shown to be using bird and raptor remains for reasons unrelated to consumption.[57] Especially interesting is the evidence of their use of feathers. Despite a detailed search, there seems to be no evidence of the behavior recorded at MSA sites in South Africa. The Chatelperronian site of Grotte du Renne has bird bones shaped with regularly spaced notches,[58] and later Aurignacian sites show the use of the radius of a griffon vulture for a flute.[59] The MSA site of Ysterfontein, South Africa, dating to between 57 and 46 kya, has numerous bird bones, but no indication of feather removal.[60] Two bone tools made from bird bones are reported from Blombos cave.[61] While some of the avifauna at Sidubu have cut marks, none has yet been reported to show the same sort of feather removal seen at Neandertal sites.[62] One of the bone tools from Sibudu is made from a bird radius.[63]

The point is that there is a behavior which is unseen in earlier hominins and also not yet recorded in the MSA, but is somewhat common at archaeological sites associated with Neandertals. Why remove feathers, especially ones that seem to have been selected on the basis of color, if not for adornment? A review of the ethnographic literature shows that the use of feathers is widespread in modern human culture, as they are costly to obtain and thus signal status.[64] If the situation were reversed, and the data suggested that Still Bay people were removing feathers, it is likely that this would be used as evidence supporting modern human behavior's origins lying in Southern Africa.

What does it mean if Neandertals show signs of symbolic thought?

1. Maybe symbolism should be removed as a sign of modern human behavior.
2. Maybe both groups evolved symbolic thought independently though parallel evolution.
3. Maybe the last common ancestor of Neandertals and modern humans had symbolic thought.
4. Maybe the species designations of Neandertals and AMH are faulty, and Neandertals should be incorporated into the *H. sapiens* lineage.

Currently, choosing between these options is difficult. Francesco d'Errico and Chris Henshilwood suggest that parallel evolution explains the evidence that Neandertals used symbolic expression, suggesting that environmental changes may have triggered the evolution of symbolism in both populations independently.[65] However, it is not clear how this is the most parsimonious explanation rather than suggesting that either the ability evolved before the two groups split (i.e., option three above) or that symbolic expression and other forms of cultural display were forming within the connections between these populations.

If the *H. helmei* hypothesis is correct, then option three works well. However, there is currently little fossil evidence to support it. In the next section, I will sketch out some reasons why the first option may be the best way forward—not in the sense that we should remove Neandertals as purveyors of modernity, but rather in the sense that the use of the word "symbolic" may be misapplied in many archaeological instances.

WHAT MAKES SOMETHING SYMBOLIC?

Recent scholarship has demonstrated that the first appearance of these indicators of modernity does not fit with traditional models of human evolution.[66] There is thus a critical need to determine what the distributions of these symbolic objects actually are, both spatially and temporally, for the totality of the Pleistocene. In the absence of this knowledge, we will be creating inferences from an incomplete record, one that is often biased toward scholarship from Western scholars and concentrated on only a few key regions rather than the entire Old World.

With this in mind, we created the Worldwide Instances of Symbolic Data Outlining Modernity (WISDOM) database, which records all known examples of human symbolic expression up till 45 kya.[67] Interested scholars can access the data and registered users can update and enter information. We currently have data from ~90 sites from almost all regions of the Old World. These data suggest that material symbolic expression can be seen in earlier, pre-AMH, hominins.[68]

While these data indicate that pre-sapiens were capable of creating objects that were previously considered to be only within the purview of modern humans, the fact that we do not know how to interpret these objects makes salient comparisons across regions and populations difficult. Perhaps the evidence of Neandertal symbolic thought can be taken as evidence that symbolism should not be seen as an indicator of modernity—or maybe that the indicators for symbolism have been misapplied.

Discussing the South African record, Chris Henshilwood and Benoît Dubreuil argue that the Blombos beads are symbols, since "in the archaeological literature beads are indisputably regarded as symbolic artifacts and indicative of 'modern' behavior."[69] They further suggest that the beads show the effects of the expansion of the temporoparietal areas of the brain.[70] As for the symbolic nature, their argument rests on two inferences. First, they note that beads are found at different levels within Blombos. Second, they argue that the mental requirements needed to be concerned with one's appearance are the same ones necessary for symbolic meaning. To support this, they note that children both tend to be unconcerned with their appearance and do not understand how the meaning of an object, symbolically speaking, depends on a collective agreement about what that object means. This behavior may be linked, in part, to neurological changes in the hominin brain. Seeing beads as evidence of symbolic thought has not gone unquestioned, with Frederick Coolidge and Thomas Wynn suggesting that perhaps they functioned as tally devices rather than as symbolically-mediated signs.[71]

Scholars have compared modern human brain function to that of chimpanzees. fMRI and other techniques have been applied to human and non-human primates in order to determine what features of the brain are unique to our species. However, this tells us little about when these changes occurred within the human lineage. Furthermore, whether or not a four-year-old is able to think symbolically cannot tell us much about what an adult early human did. Just as ontogeny does not recapitulate phylogeny, the timing of developmental stages in *Homo sapiens* does not necessarily reflect the evolutionary history of our species.

Another reason for possibly removing symbolism as an indicator of modern behavior comes from non-human primate and animal studies. William McGrew draws attention to primate behaviors, such as male chimpanzees tearing leaves to draw attention from females and the hand-clasps between grooming partners, which have a high level of variation within a group.[72] McGrew has also argued that the hand-clasp could be a group identifier.[73] Interestingly, he notes that if such behaviors were noted in human groups, they would easily be interpreted as ethnic markers. Other examples include the drumming behavior of Tai chimpanzees, in which the drumming is argued to be a way to communicate position in the forest.[74] Christophe Boesch has argued that one chimp in particular used this system to communicate to others that he was either changing direction or going to take a rest. Boesch suggests that such a system is "symbolic." Similarly, Goodall's description of a rain dance and of "waterfall displays" suggests some sort of deeper meaning to these acts.[75]

Yet, we need to be careful. If a chimpanzee picks up a rock and carries it to a nearby tree, that behavior could make the cover of *National Geographic*.[76] Would a similar behavior, seen in a monkey, receive the same attention? Furthermore, we need to be careful when comparing behaviors of modern chimps to those of early humans. Chimps are very good at being chimps, but not so good at being humans. And the same is true of earlier hominins. They may not have been good at being AMH, but they excelled at their own behaviors.

The relevance of the term "symbolic" and how it can be best applied to the paleoanthropological record has also been of recent concern.[77] Terrence Deacon has had the most influence on anthropologists in terms of the emphasis on symbol-making, especially with his emphasis on Peircean semiotics.[78] Peirce saw signs as composed of three related components: the sign-vehicle, the object, and the interpretant.[79] Under a Peircean system one can examine the sign itself, how the sign is related to its object, and how the sign is related to its interpretant. The most famous of these relationships, at least for archaeologists, is that between a sign-vehicle and its object, which is where Peirce's Icon-Index-Symbol distinction comes from. It is his definition—namely that a symbolic sign is one in which the ground between the object and the sign is via conventional agreement (rather than through resemblance or causal relation)—that archaeologists tend to apply to the past.

But this distinction is difficult to apply to the remote past since we often do not know the cultural context in which these artifacts were produced.[80] One way around this is to apply another of Pierce's trichotomies, one which looks at how the sign itself functions. Peirce defined three types of sign vehicles: *qualisigns, sinsigns,* and *legisigns*. Qualisigns, like icons, are derived from qualities. What makes a qualisign confusing is that it does not signify anything except as it is embodied in an object (like the color blue, which can be embodied in a picture or piece of fabric). The second type of sign-vehicle is the *sinsign,* a sign which contains several qualisigns. Finally, a *legisign* is when the sign vehicle signifies based on convention. Replicas are an individual instance of a legisign, which makes them a special category of sinsigns where their significance is based both on being a replica of a legisign and on the features of their occurrence.

Perhaps by thinking in these terms, we can remove the fetishization of the symbol as the thing that makes us human. Using the qualisign/sinsign/legisign approach allows for the identification of a purposefully created system without the assumption of a symbolic referent. For example, engraved lines are found in different instances, such as on ostrich eggshell, ochre, bone, and stone. Early humans were embedding qualisigns in these materials. We can perhaps trace qualisigns through different media, such as noting the persistence of the # symbol in disparate archaeological assemblages. Perhaps these are examples of humans copying phosphenes.[81] Under this system, most of the early examples recorded in the WISDOM database are sinsigns. They show humans interacting with the natural world and creating artifacts, but not necessarily for the consumption of others. It may have had meaning to an individual, but we cannot tell if it was part of a larger overarching system of belief. However, at sites with many engraved eggshells, we can infer, if enough information is present, that a legisign existed. Thus, in these cases the individuals making these artifacts were involved in a larger system in which they hoped to produce some effect on their community members' minds.

CONCLUSION

Deacon's hypothesis linking symbolic thought, language, and human uniqueness has influenced much of the work in this area, with archaeologists searching for symbolic behavior in the past though the use of the material record. With some notable exceptions, archaeological signs of modernity tend to postdate 300,000 years ago. If we make the assumption that humans before 300 kya were not behaving in a modern way, what can this tell us about human evolution in general? The suggestion that a single genetic change is the main cause of behavioral modernity is untenable for a number of reasons, not the least of which is that there is no gene for behavioral modernity. Even a suite of genetic changes is unable to explain the behavioral shifts behind the origins of modern human behavior, as these are the result of a complex interplay between various evolutionary forces and inheritance systems.[82]

The necessarily complex answer no doubt entails not just the use of anthropological data, but the incorporation of theories and methodologies from fields such as neuroscience, information theory, semiotics, innovation theory, and genomics. Kim Sterelny's Apprentice Theory[83] and Michael Tomasello's Joint and Collective Intentionality[84] models can provide insight into this time period, though these models have yet to be applied directly to the question of modern human origins. An evolutionary approach to human wisdom can incorporate all these models.

So maybe wisdom stands a chance in the anthropological literature. Wisdom can be seen as the central aspect in the complex decision-making necessary when successfully navigating social networks and dynamic niches. We can then center questions not on what these objects meant to early humans, but on *how* they were able to mean something, through the application of Peircean semiotics. This ability to create objects that not only have meaning, but that are created with the intent to produce a specific meaning/response in the mind of another person, is a critical part of human behavior.

NOTES

1. See Susan Blum's comment in Marc Kissel and Agustín Fuentes, "From Hominid to Human: The Role of Human Wisdom and Distinctiveness in the Evolution of Modern Humans," *Philosophy, Theology and the Sciences* 3 (2016): 235–36.

2. Thierry Hoquet, "Biologization of Race and Racialization of the Human: Bernier, Buffon, Linnaeus," in *The Invention of Race: Scientific and Popular Representation*, ed. Nicolas Bancel, Thomas David, and Dominic Thomas (New York: Routledge, 2014), 17–32.

3. Keith W. Kintigh et al., "Grand Challenges for Archaeology," *Proceedings of the National Academy of Sciences* 111 (2014): 879–80; Sally McBrearty, "Down with the Revolution," in *Rethinking the Human Revolution: New Behavioural and Biological Perspectives on the Origin and Dispersal of Modern Humans*, ed. Paul Mellars, Katie Boyle, Ofer Bar-Yosef, and Chris Stringer (Cambridge: McDonald Institute for Archaeological Research, 2007), 133–51; Lee R. Berger et al., "A New Species of Homo from the Dinaledi Chamber, South Africa," *Elife* 4 (2015): e09560.

4. Gilliane F. Monnier, "The Lower/Middle Paleolithic Periodization in Western Europe: An Evaluation," *Current Anthropology* 47 (2006): 709–44.

5. Paul Mellars, "The Impossible Coincidence: A Single-Species Model for the Origins of Modern Human Behavior in Europe," *Evolutionary Anthropology* 14 (2005): 12–27; Paul Mellars, "Major Issues in the Emergence of Modern Humans," *Current Anthropology* 30 (1989): 349–85.

6. Colin Renfrew, "The Sapient Behaviour Paradox: How to Test for Potential?" in *Modelling the Early Human Mind*, ed. Paul Mellars and K. Gibson (Cambridge: McDonald Institute for Archaeological Research, 1996), 11–15.

7. Richard G. Klein, "The Archeology of Modern Human Origins," *Evolutionary Anthropology* 1 (1992): 5–14; Sally Mcbrearty and Alison Brooks, "The Revolution That Wasn't: A New Interpretation of the Origin of Modern Human Behavior," *Journal of Human Evolution* 39 (2000): 453–563; Francesco d'Errico, "The Invisible Frontier: A Multi-Species Model for the Origin of Behavioral Modernity," *Evolutionary Anthropology* 12 (2003): 188–202; Christopher S. Henshilwood and Curtis Marean, "The Origin of Modern Human Behavior," *Current Anthropology* 44 (2003): 627–51.

8. See J. Lind, P. Lindenfors, S. Ghirlanda, K. Liden, and M. Enquist, "Dating Human Cultural Capacity Using Phylogenetic Principles," *Nature Scientific Reports* 3 (2013): 1–5.

9. John J. Shea, "Homo Sapiens Is as Homo Sapiens Was," *Current Anthropology* 52 (2011): 1–35.

10. What is meant by "modern" is never clear. See Shea, ibid., for an overview of some of the problems with this term. In general, paleoanthropologists use "modern" to refer to the hypothesis that there is something inherently different in the behavior of modern humans compared to the first *H. sapiens*.

11. See Dylan Belton, "The Human as World-Open Spirit," this volume.

12. Richard G. Klein, "Archeology and the Evolution of Human Behavior," *Evolutionary Anthropology* 9 (2000): 17–18.

13. Richard G. Klein and Blake Edgar, *The Dawn of Human Culture* (New York: Wiley, 2002).

14. Wolfgang Enard et al., "Molecular Evolution of FOXP2, a Gene Involved in Speech and Language," *Nature* 418 (2002): 869–72.

15. Francesco d'Errico et al., "From the Origin of Language to the Diversification of Languages: What Can Archaeology and Palaeoanthropology Say?" in *Becoming Eloquent: Advances in the Emergence of Language*, ed. Francesco d'Errico and Jean-Marie Hombert (Amsterdam: John Benjamins, 2009), 13–68.

16. Johannes Krause et al., "The Derived FOXP2 Variant of Modern Humans Was Shared with Neandertals," *Current Biology* 17.21 (2007): 1908–12; Graham Coop et al., "The Timing of Selection at the Human FOXP2 Gene," *Molecular Biology and Evolution* 25 (2008): 1257–59.

17. Coop et al., "Timing of Selection."

18. Tomislav Maricic et al., "A Recent Evolutionary Change Affects a Regulatory Element in the Human FOXP2 Gene," *Molecular Biology and Evolution* 30.4 (2013): 844–52.

19. John Hawks, Eric T. Wang, Gregory M. Cochran, Henry C. Harpending, and Robert K. Moyzis, "Recent Acceleration of Human Adaptive Evolution," *Proceedings of the National Academy of Sciences* 104.52 (2007): 20753–58.

20. Ian McDougall, Francis H. Brown, and John G. Fleagle, "Stratigraphic Placement and Age of Modern Humans from Kibish, Ethiopia," *Nature* 433 (2005): 733–36.

21. Tim D. White, Berhane Asfaw, David DeGusta, Henry Gilbert, Gary D. Richards, Gen Suwa, and F. Clark Howell, "Pleistocene Homo Sapiens from Middle Awash, Ethiopia," *Nature* 423 (2003): 742–47.

22. Jeffrey H. Schwartz and Ian Tattersall, "Fossil Evidence for the Origin of Homo Sapiens," *American Journal of Physical Anthropology* 143.S51 (2010): 94–121.

23. Milford Wolpoff, Alan Thorne, Jan Jelínek, and Zhang Yinyun, "The Case for Sinking Homo Erectus: 100 Years of Pithecanthropus Is Enough!" in *100 Years of Pithecanthropus: The* Homo Erectus *Problem*, ed. Jens Lorenz Franzen (Frankfurt: Courier Forschungsinstitut Senckenberg, 1994), 341–61.

24. P. V. Tobias, "The Bearing of Fossils and Mitochondrial DNA on the Evolution of Modern Humans, with a Critique of the 'Mitochondrial Eve' Hypothesis," *South African Archaeological Bulletin* 50 (1995): 158.
25. Rachel Caspari and Milford H. Wolpoff, "The Process of Modern Human Origins: The Evolutionary and Demographic Changes Giving Rise to Modern Humans," in *Origins of Modern Humans: Biology Reconsidered,* ed. Fred Smith (New York: Wiley, 2013), 355–90.
26. Glynn LL. Isaac, "Sorting Out the Muddle in the Middle: An Anthropologist's Post-Conference Appraisal," in *After the Australopithecines: Stratigraphy, Ecology and Culture Change in the Middle Pleistocene,* ed. Karl W. Butzer and Glynn LL. Isaac (The Hague: Mouton, 1975), 875–87.
27. Marta Lahr and Robert Foley, "Multiple Dispersals and Modern Human Origins," *Evolutionary Anthropology* 3 (1994): 48–60; Marta Lahr and Robert Foley, "Towards a Theory of Modern Human Origins: Geography, Demography, and Diversity in Recent Human Evolution," *Yearbook of Physical Anthropology* 41 (1998): 137–76; Marta Lahr and Robert Foley, "Mode 3, Homo Helmei, and the Pattern of Human Evolution in the Middle Pleistocene," in *Human Roots: Africa and Asia in the Middle Pleistocene.,* ed. K. Robson Brown (Bristol: Western Academic & Specialist Press, 2001), 23–39; Robert Foley and Marta Lahr, "Mode 3 Technologies and the Evolution of Modern Humans," *Cambridge Archaeological Journal* 7 (1997): 3–36; Robert Foley and Marta Lahr, "On Stony Ground: Lithic Technology, Human Evolution, and the Emergence of Culture," *Evolutionary Anthropology* 12 (2003): 109–22.
28. Günter Brauer et al., "Pathological Alterations in the Archaic Homo Sapiens Cranium from Eliye Springs, Kenya," *American Journal of Physical Anthropology* 120 (2003): 200–4.
29. A hypodigm is the corpus of material from which a researcher can make inferences about a species or population.
30. J. M. Bermúdez de Castro et al., "The Atapuerca Sites and Their Contribution to the Knowledge of Human Evolution in Europe," *Evolutionary Anthropology* 13 (2004): 25–41.
31. Grahame Clark, *World Prehistory in New Perspective* (Cambridge: Cambridge University Press, 1977).
32. James L. Boone and Eric A. Smith, "Is It Evolution Yet? A Critique of Evolutionary Archaeology," *Current Anthropology* 39 (1998): S141–S173; Stephen Shennan, "Population, Culture History, and the Dynamcs of Culture Change," *Current Anthropology* 41 (2000): 811–33; Stephen Shennan and J. R. Wilkinson, "Ceramic Style Change and Neutral Evolution: A Case Study from Neolithic Europe," *American Antiquity* 66.4 (2001): 577–93.
33. Thomas Wynn and Forrest Tierson, "Regional Comparison of the Shapes of Later Acheulean Handaxes," *American Antiquity* 92 (1990): 73–84.
34. Curtis W. Marean, "An Evolutionary Anthropological Perspective on Modern Human Origins," *Annual Review of Anthropology* 44 (2015): 533–56.
35. João Zilhão et al., "Symbolic Use of Marine Shells and Mineral Pigments by Iberian Neandertals," *Proceedings of the National Academy of Sciences* 107 (2010): 1023–28; Marco Peresani et al., "An Ochered Fossil Marine Shell From the Mousterian of Fumane Cave, Italy," *PLOS ONE* 8 (2013): 1–15; Wil Roebroeks et al., "Use of Red Ochre by Early Neandertals," *Proceedings of the National Academy of Sciences* 109 (2012): 1889–94.
36. V. Gordon Childe, "The Urban Revolution," *The Town Planning Review* 21 (1950): 3–17.
37. Bruce Trigger, *A History of Archaeological Thought* (Cambridge: Cambridge University Press, 2006).
38. J. Mark Kenoyer, "The Harappan State: Was It or Wasn't It?" *Wisconsin Archaeological Reports* 3 (1994): 71–80.
39. Leslie White, "The Symbol: The Origin and Basis of Human Behavior," *Philosophy of Science* 7 (1940): 451.

40. Ibid., 453.

41. Terry Hopkinson, "'Man the Symboller': A Contemporary Origins Myth," *Archaeological Dialogues* 20 (2013): 215–41.

42. See, for example, Mellars, "The Impossible Coincidence"; Klein and Edgar, *The Dawn of Human Culture*.

43. Alex Mackay and Aara Welz, "Engraved Ochre from a Middle Stone Age Context at Klein Kliphuis in the Western Cape of South Africa," *Journal of Archaeological Science* 35 (2008): 1521–32; Christopher S. Henshilwood, Francesco d'Errico, and Ian Watts, "Engraved Ochers from the Middle Stone Age Levels at Blombos Cave, South Africa," *Journal of Human Evolution* 57 (2009): 27–47; Francesco d'Errico, Renata García Moreno, and Riaan F. Rifkin, "Technological, Elemental and Colorimetric Analysis of an Engraved Ochre Fragment from the Middle Stone Age Levels of Klasies River Cave 1, South Africa," *Journal of Archaeological Science* 39 (2012): 942–52.

44. Christopher Henshilwood, Francesco d'Errico, Marian Vanhaeren, Karen van Niekerk, and Zenobia Jacobs, "Middle Stone Age Shell Beads from South Africa," *Science* 304.5669 (2004): 404; Francesco d'Errico, Christopher Henshilwood, Martin Vanhaeren, and Karen van Niekerk, "Nassarius Kraussianus Shell Beads from Blombos Cave: Evidence for Symbolic Behaviour in the Middle Stone Age," *Journal of Human Evolution* 48 (2005): 3–24; Abdeljalil Bouzouggar et al., "82,000-Year-Old Shell Beads from North Africa and Implications for the Origins of Modern Human Behavior," *Proceedings of the National Academy of Sciences* 104 (2007): 9964–69; Rudolf Botha, "Prehistoric Shell Beads as a Window on Language Evolution," *Language & Communication* 28 (2008): 197–212.

45. Sabine Gaudzinski, "Middle Palaeolithic Bone Tools from the Open-Air Site Salzgitter-Lebenstedt (Germany)," *Journal of Archaeological Science* 26.2 (1999): 125–41; P. P. A. Mazza et al., "A New Palaeolithic Discovery: Tar-Hafted Stone Tools in a European Mid-Pleistocene Bone-Bearing Bed," *Journal of Archaeological Science* 33 (2006): 1310–18; Lucinda Backwell, Francesco d'Errico, and Lyn Wadley, "Middle Stone Age Bone Tools from the Howiesons Poort Layers, Sibudu Cave, South Africa," *Journal of Archaeological Science* 35 (2008): 1566–80; Marie Soressi et al., "Neandertals Made the First Specialized Bone Tools in Europe," *Proceedings of the National Academy of Sciences* 110.35 (2013): 14186–90.

46. Marlize Lombard, "Evidence of Hunting and Hafting during the Middle Stone Age at Sibidu Cave, KwaZulu-Natal, South Africa: A Multianalytical Approach," *Journal of Human Evolution* 48 (2005): 279–300; Armelle Charrié-Duhaut et al., "First Molecular Identification of a Hafting Adhesive in the Late Howiesons Poort at Diepkloof Rock Shelter (Western Cape, South Africa)," *Journal of Archaeological Science* 40 (2013): 3506–18; Paola Villa, Anne Delagnes, and Lyn Wadley, "A Late Middle Stone Age Artifact Assemblage from Sibudu (KwaZulu-Natal): Comparisons with the European Middle Paleolithic," *Journal of Archaeological Science* 32 (2005): 399–422; Lyn Wadley, "Recognizing Complex Cognition through Innovative Technology in Stone Age and Palaeolithic Sites," *Cambridge Archaeological Journal* 23 (2013): 163–83.

47. Paul A. Mellars, "The Character of the Middle-Upper Palaeolithic Transition in South-West France," in *The Explanation of Culture Change: Models in Prehistory*, ed. Colin Renfrew (London: Duckworth, 1973), 255–76.

48. Sally McBrearty and Alison Brooks, "The Revolution That Wasn't: A New Interpretation of the Origin of Modern Human Behavior," *Journal of Human Evolution* 39 (2000): 453–563.

49. Christopher S. Henshilwood and Curtis Marean, "The Origin of Modern Human Behavior," *Current Anthropology* 44 (2003): 627–51.

50. Trenton W. Holliday, "Comment on 'The Origin of Modern Human Behavior,'" *Current Anthropology* 44 (2003): 639–40.

51. Perhaps instead wolves were able to exploit an open niche in a way that the coyotes could not.

52. Erik Trinkaus and Pat Shipman, *The Neandertals: Changing the Image of Mankind*, 1st ed. (New York: Knopf, 1992).

53. Bernard Vandermeersch and María Dolores Garralda, "Neanderthal Geographical and Chronological Variation," in *Continuity and Discontinuity in the Peopling of Europe: One Hundred Fifty Years of Neanderthal Study*, ed. Silvana Condemi and Gerd-Christian Weniger (New York: Springer, 2011), 113–25.

54. Richard E. Green et al., "A Draft Sequence of the Neandertal Genome," *Science* 328 (2010): 710–22.

55. Zilhão et al., "Symbolic Use of Marine Shells."

56. A. W. G. Pike et al., "U-Series Dating of Paleolithic Art in 11 Caves in Spain," *Science* 336 (2012): 1409–13.

57. Clive Finlayson et al., "Birds of a Feather: Neanderthal Exploitation of Raptors and Corvids," *PlOS ONE* 7 (2012): e45927; Eugène Morin and Véronique Laroulandie, "Presumed Symbolic Use of Diurnal Raptors by Neanderthals," *PLOS ONE* 7 (2012): 1–5; Davorka Radovčić, Ankica Oros Sršen, Jakov Radovčić, and David W. Frayer, "Evidence for Neandertal Jewelry: Modified White-Tailed Eagle Claws at Krapina," *PLOS ONE* 10 (2015): e0119802.

58. Francesco d'Errico, João Zilhão, Michèle Julien, Dominique Baffier, and Jacques Pelegrin, "Neanderthal Acculturation in Western Europe? A Critical Review of the Evidence and Its Interpretation," *Current Anthropology* 39.S1 (1998): 1–44.

59. Nicholas J. Conard, Maria Malina, and Susanne C. Münzel, "New Flutes Document the Earliest Musical Tradition in Southwestern Germany," *Nature* 460 (2009): 737–40.

60. Richard G. Klein et al., "The Ysterfontein 1 Middle Stone Age Site, South Africa, and Early Human Exploitation of Coastal Resources," *Proceedings of the National Academy of Sciences* 101 (2004): 5708–15.

61. Francesco d'Errico and Christopher S. Henshilwood, "Additional Evidence for Bone Technology in the Southern African Middle Stone Age," *Journal of Human Evolution* 52 (2007): 142–63.

62. Aurore Val, "New Data on the Avifauna from the Middle Stone Age Layers of Sibudu Cave, South Africa: Taphonomic and Palaeoenvironmental Implications," *Quaternary International* 421 (2016): 173–89.

63. Francesco d'Errico, Lucinda R. Backwell, and Lyn Wadley, "Identifying Regional Variability in Middle Stone Age Bone Technology: The Case of Sibudu Cave," *Journal of Archaeological Science* 39 (2012): 2479–95.

64. Finlayson et al., "Birds of a Feather."

65. Francesco d'Errico and Christopher S. Henshilwood, "The Origin of Symbolically Mediated Behaviour: From Antagonistic Scenarios to a Unified Research Strategy," in *Homo Symbolicus: The Dawn of Language, Imagination and Spirituality*, ed. Francesco d'Errico and Christopher S. Henshilwood (Amsterdam: John Benjamins, 2011), 49–74.

66. Marc Kissel and Agustín Fuentes, "From Hominid to Human: The Role of Human Wisdom and Distinctiveness in the Evolution of Modern Humans," *Philosophy, Theology and the Sciences* 3 (2016): 217–44.

67. See https://marckissel.shinyapps.io/wisdom_database/.

68. Marc Kissel and Agustín Fuentes, "Semiosis in the Pleistocene," *Cambridge Archaeological Journal* 3.21 (2017): 1–16.

69. Christopher S. Henshilwood and Benoît Dubreuil, "Reading the Artifacts: Gleaning Language Skills from the Middle Stone Age in Southern Africa," in *The Cradle of Language*, ed. Rudolf Botha and Chris Knight (Oxford: Oxford University Press, 2009), 50.

70. Christopher S. Henshilwood and Benoît Dubreuil, "The Still Bay and Howiesons Poort, 77–59 Ka," *Current Anthropology* 52 (2011): 361–400.

71. Frederick L. Coolidge and Thomas Wynn, "Comment on 'The Still Bay and Howiesons Poort, 77–59 Ka: Symbolic Material Culture and the Evolution of the Mind during the African Middle Stone Age,'" *Current Anthropology* 52 (2011): 380–82.

72. William C. McGrew, "Pan Symbolicus: A Cultural Primatologist's Viewpoint," in *Homo Symbolicus: The Dawn of Language, Imagination and Spirituality*, ed. Christopher S. Henshilwood and Francesco d'Errico (Amsterdam: John Benjamins, 2011), 1–12.

73. William C. McGrew, *The Cultured Chimpanzee: Reflections on Cultural Primatology* (Cambridge: Cambridge University Press, 2004).

74. Christophe Boesch, "Symbolic Communication in Wild Chimpanzees?" *Human Evolution* 6 (1991): 81–89.

75. Jane Goodall, *The Chimpanzees of Gombe: Patterns of Behavior* (Cambridge, MA: Belknap Press, 1986).

76. See, for example, Hjalmar Kühl et al., "Chimpanzee Accumulative Stone Throwing," *Nature Scientific Reports* 6 (2016): 1–8.

77. Duilio Garofoli, "Cognitive Archaeology without Behavioral Modernity: An Eliminativist Attempt," *Quaternary International* 405 (2015): 125–35.

78. Terrence W. Deacon, *The Symbolic Species: The Co-Evolution of Language and the Brain* (New York: W. W. Norton, 1997).

79. Charles Peirce, *The Essential Peirce: Selected Philosophical Writings, Vol. 1,* ed. E. Houser and C. Kloesel (Bloomington: Indiana Univeristy Press, 1992).

80. Kissel and Fuentes, "Semiosis in the Pleistocene."

81. Derek Hodgson, "Decoding the Blombos Engravings, Shell Beads and Diepkloof Ostrich Eggshell Patterns," *Cambridge Archaeological Journal* 24 (2014): 57–69.

82. Eva Jablonka and Marion J. Lamb, *Evolution in Four Dimensions: Genetic, Epigenetic, Behavioral, and Symbolic Variation in the History of Life* (Cambridge, MA: MIT Press, 2005).

83. Kim Sterelny, *The Evolved Apprentice: How Evolution Made Humans Unique* (Cambridge, MA: MIT Press, 2012).

84. Michael Tomasello, *A Natural History of Human Thinking* (Cambridge, MA: Harvard University Press, 2014).

BIBLIOGRAPHY

Backwell, Lucinda, Francesco d'Errico, and Lyn Wadley. "Middle Stone Age Bone Tools from the Howiesons Poort Layers, Sibudu Cave, South Africa." *Journal of Archaeological Science* 35 (2008): 1566–80.

Berger, Lee R., John Hawkes, D. J. de Ruiter, Steven E. Churchill et al. "A New Species of Homo from the Dinaledi Chamber, South Africa." *Elife* 4 (2015): e09560.

Bermúdez de Castro, J. M., M. Martinón-Torres, E. Carbonell et al. "The Atapuerca Sites and Their Contribution to the Knowledge of Human Evolution in Europe." *Evolutionary Anthropology* 13 (2004): 25–41.

Boesch, Christophe. "Symbolic Communication in Wild Chimpanzees?" *Journal of Human Evolution* 6 (1991): 81–89.

Boone, James L., and Eric A. Smith. "Is It Evolution Yet? A Critique of Evolutionary Archaeology." *Current Anthropology* 39 (1998): S141–S173.

Botha, Rudolf. "Prehistoric Shell Beads as a Window on Language Evolution." *Language & Communication* 28 (2008): 197–212.

Bouzouggar, Abdeljalil, Nick Barton, Marian Vanhaerend et al. "82,000-Year-Old Shell Beads from North Africa and Implications for the Origins of Modern Human Behavior." *Proceedings of the National Academy of Sciences* 104 (2007): 9964–69.

Brauer, G., C. Gorden, G. Delling et al. "Pathological Alterations in the Archaic Homo Sapiens Cranium from Eliye Springs, Kenya." *American Journal of Physical Anthropology* 120 (2003): 200–4.

Caspari, Rachel, and Milford H. Wolpoff. "The Process of Modern Human Origins: The Evolutionary and Demographic Changes Giving Rise to Modern Humans." In *Origins of Modern Humans: Biology Reconsidered,* edited by Fred Smith, 355–90. New York: Wiley, 2013.

Charrié-Duhaut, Armelle, Guillaume Porraz, Caroline R. Cartwright et al. "First Molecular Identification of a Hafting Adhesive in the Late Howiesons Poort at Diepkloof Rock Shelter (Western Cape, South Africa)." *Journal of Archaeological Science* 40 (2013): 3506–18.

Childe, V. Gordon. "The Urban Revolution." *The Town Planning Review* 21 (1950): 3–17.

Clark, Grahame. *World Prehistory in New Perspective*. New York: Cambridge University Press, 1977.

Conard, Nicholas J., Maria Malina, and Susanne C. Münzel. "New Flutes Document the Earliest Musical Tradition in Southwestern Germany." *Nature* 460 (2009): 737–40.

Coolidge, Frederick L., and Thomas Wynn. "Comment on 'The Still Bay and Howiesons Poort, 77–59 Ka: Symbolic Material Culture and the Evolution of the Mind during the African Middle Stone Age.'" *Current Anthropology* 52 (2011): 380–82.

Coop, Graham, Kevin Bullaughey, Francesca Luca, and Molly Przeworski. "The Timing of Selection at the Human FOXP2 Gene." *Molecular Biology and Evolution* 25 (2008): 1257–59.

Deacon, Terrence W. *The Symbolic Species: The Co-Evolution of Language and the Brain*. New York: W. W. Norton, 1997.

Enard, Wolfgang, Molly Przeworski, Simon E. Fisher et al. "Molecular Evolution of FOXP2, a Gene Involved in Speech and Language." *Nature* 418 (2002): 869–72.

d'Errico, Francesco. "The Invisible Frontier: A Multi-Species Model for the Origin of Behavioral Modernity." *Evolutionary Anthropology* 12 (2003): 188–202.

d'Errico, Francesco, and Christopher S. Henshilwood. "Additional Evidence for Bone Technology in the Southern African Middle Stone Age." *Journal of Human Evolution* 52 (2007): 142–63.

———. "The Origin of Symbolically Mediated Behaviour: From Antagonistic Scenarios to a Unified Research Strategy." In *Homo Symbolicus: The Dawn of Language, Imagination and Spirituality*, edited by Francesco d'Errico and Christopher S. Henshilwood, 49–74. Amsterdam: John Benjamins, 2011.

d'Errico, Francesco, Christopher S. Henshilwood, Marian Vanhaeren, and Karen van Niekerk. "Nassarius Kraussianus Shell Beads from Blombos Cave: Evidence for Symbolic Behaviour in the Middle Stone Age." *Journal of Human Evolution* 48 (2005): 3–24.

d'Errico, Francesco, Lucinda R. Backwell, and Lyn Wadley. "Identifying Regional Variability in Middle Stone Age Bone Technology: The Case of Sibudu Cave." *Journal of Archaeological Science* 39 (2012): 2479–95.

d'Errico, Francesco, João Zilhão, Michele Julien, Dominique Baffier, and Jacques Pelegrin. "Neanderthal Acculturation in Western Europe? A Critical Review of the Evidence and Its Interpretation." *Current Anthropology* 39.S1 (1998): 1–44.

d'Errico, Francesco, Marian Vanhaeren, Christopher S. Henshilwood, et al. "From the Origin of Language to the Diversification of Languages: What Can Archaeology and Palaeoanthropology Say?" In *Becoming Eloquent: Advances in the Emergence of Language*, edited by Francesco d'Errico and Jean-Marie Hombert, 13–68. Amsterdam: John Benjamins, 2009.

d'Errico, Francesco, Renata García Moreno, and Riaan F. Rifkin. "Technological, Elemental and Colorimetric Analysis of an Engraved Ochre Fragment from the Middle Stone Age Levels of Klasies River Cave 1, South Africa." *Journal of Archaeological Science* 39 (2012): 942–52.

Finlayson, Clive, Kimberly Brown, Ruth Blasco et al. "Birds of a Feather: Neanderthal Exploitation of Raptors and Corvids." *PLOS ONE* 7 (2012): e45927.

Foley, Robert, and Marta Mirazon Lahr. "Mode 3 Technologies and the Evolution of Modern Humans." *Cambridge Archaeological Journal* 7 (1997): 3–36.

———. "On Stony Ground: Lithic Technology, Human Evolution, and the Emergence of Culture." *Evolutionary Anthropology* 12 (2003): 109–22.

Garofoli, Duilio. "Cognitive Archaeology without Behavioral Modernity: An Eliminativist Attempt." *Quaternary International* 405 (2015): 125–35.

Gaudzinski, Sabine. "Middle Palaeolithic Bone Tools from the Open-Air Site Salzgitter-Lebenstedt (Germany)." *Journal of Archaeological Science* 26.2 (1999): 125–41.

Goodall, Jane. *The Chimpanzees of Gombe: Patterns of Behavior*. Cambridge, MA: Belknap Press, 1986.

Green, R. E., J. Krause, A. W. Briggs et al. "A Draft Sequence of the Neandertal Genome." *Science* 328 (2010): 710–22.

Hawks, John, Eric T. Wang, Gregory M. Cochran, Henry C. Harpending, and Robert K. Moyzis. "Recent Acceleration of Human Adaptive Evolution." *Proceedings of the National Academy of Sciences* 104.52 (2007): 20753–58.

Henshilwood, Christopher S., and Benoît Dubreuil. "Reading the Artifacts: Gleaning Language Skills from the Middle Stone Age in Southern Africa." In *The Cradle of Language*, edited by Rudolf Botha and Chris Knight, 41–61. Oxford: Oxford University Press, 2009.

———. "The Still Bay and Howiesons Poort, 77–59 Ka." *Current Anthropology* 52 (2011): 361–400.

Henshilwood, Christopher S., and Curtis Marean. "The Origin of Modern Human Behavior." *Current Anthropology* 44 (2003): 627–51.

Henshilwood, Christopher S, Francesco d'Errico, and Ian Watts. "Engraved Ochers from the Middle Stone Age Levels at Blombos Cave, South Africa." *Journal of Human Evolution* 57 (2009): 27–47.

Henshilwood, Christopher S., Francesco d'Errico, Marian Vanhaeren, Karen Van Niekerk, and Zenobia Jacobs. "Middle Stone Age Shell Beads from South Africa." *Science* 304.5669 (2004): 404.

Hodgson, Derek. "Decoding the Blombos Engravings, Shell Beads and Diepkloof Ostrich Eggshell Patterns." *Cambridge Archaeological Journal* 24 (2014): 57–69.

Holliday, Trenton W. "Comment on 'The Origin of Modern Human Behavior.'" *Current Anthropology* 44 (2003): 639–40.

Hopkinson, Terry. "'Man the Symboller': A Contemporary Origins Myth." *Archaeological Dialogues* 20 (2013): 215–41.

Hoquet, Thierry. "Biologization of Race and Racialization of the Human: Bernier, Buffon, Linnaeus." In *The Invention of Race: Scientific and Popular Representation*, edited by Nicolas Bancel, Thomas David, and Dominic Thomas, 17–32. New York: Routledge, 2014.

Isaac, Glynn LL. "Sorting Out the Muddle in the Middle: An Anthropologist's Post-Conference Appraisal." In *After the Australopithecines: Stratigraphy, Ecology and Culture Change in the Middle Pleistocene*, edited by Karl W. Butzer and Glynn LL. Isaac, 875–87. The Hague: Mouton, 1975.

Jablonka, Eva, and Marion J. Lamb. *Evolution in Four Dimensions: Genetic, Epigenetic, Behavioral, and Symbolic Variation in the History of Life*. Cambridge, MA: MIT Press, 2005.

Kenoyer, J. Mark. "The Harappan State: Was It or Wasn't It?" *Wisconsin Archaeological Reports* 3 (1994): 71–80.

Kintigh, Keith W., Jeffrey H. Altschul, Mary C. Beaudry et al. "Grand Challenges for Archaeology." *Proceedings of the National Academy of Sciences* 111 (2014): 879–80.

Kissel, Marc, and Agustín Fuentes. "From Hominid to Human: The Role of Human Wisdom and Distinctiveness in the Evolution of Modern Humans." *Philosophy, Theology and the Sciences* 3 (2016): 217–44.

———. "Semiosis in the Pleistocene." *Cambridge Archaeological Journal* 3.21 (2017): 1–16.

Klein, Richard G. "Archeology and the Evolution of Human Behavior." *Evolutionary Anthropology* 9 (2000): 17–36.

———. 'The Archeology of Modern Human Origins.' *Evolutionary Anthropology* 1 (1992): 5–14.

Klein, Richard G., and Blake Edgar. *The Dawn of Human Culture*. New York: Wiley, 2002.

Klein, Richard G., Graham Avery, Kathryn Cruz-Uribe et al. "The Ysterfontein 1 Middle Stone Age Site, South Africa, and Early Human Exploitation of Coastal Resources." *Proceedings of the National Academy of Sciences* 101 (2004): 5708–15.

Krause, Johannes, Carles Lalueza-Fox, Ludovic Orlando et al. "The Derived FOXP2 Variant of Modern Humans Was Shared with Neandertals." *Current Biology* 17.21 (2007): 1908–12.

Lahr, Marta Mirazon, and Robert Foley. "Mode 3, Homo Helmei, and the Pattern of Human Evolution in the Middle Pleistocene." In *Human Roots: Africa and Asia in the Middle Pleistocene*, edited by K. Robson Brown, 23–39. Bristol: Western Academic & Specialist Press, 2001.

———. "Towards a Theory of Modern Human Origins: Geography, Demography, and Diversity in Recent Human Evolution." *Yearbook of Physical Anthropology* 41 (1998): 137–76.

Lahr, Marta, and Richard Foley. "Multiple Dispersals and Modern Human Origins." *Evolutionary Anthropology* 3 (1994): 48–60.

Lombard, Marlize. "Evidence of Hunting and Hafting during the Middle Stone Age at Sibidu Cave, KwaZulu-Natal, South Africa: A Multianalytical Approach." *Journal of Human Evolution* 48 (2005): 279–300.

Mackay, Alex, and Aara Welz. "Engraved Ochre from a Middle Stone Age Context at Klein Kliphuis in the Western Cape of South Africa." *Journal of Archaeological Science* 35 (2008): 1521–32.

Marean, Curtis W. "An Evolutionary Anthropological Perspective on Modern Human Origins." *Annual Review of Anthropology* 44 (2015): 533–56.

Maricic, Tomislav, Viola Gu, Oleg Georgiev et al. "A Recent Evolutionary Change Affects a Regulatory Element in the Human FOXP2 Gene." *Molecular Biology and Evolution* 30.4 (2013): 844–52.

Mazza, P. P. A., Fabio Martini, Benedetto Sala, et al. "A New Palaeolithic Discovery: Tar-Hafted Stone Tools in a European Mid-Pleistocene Bone-Bearing Bed." *Journal of Archaeological Science* 33 (2006): 1310–18.

McBrearty, Sally. "Down with the Revolution." In *Rethinking the Human Revolution: New Behavioural and Biological Perspectives on the Origin and Dispersal of Modern Humans*, edited by Paul Mellars, Katie Boyle, Ofer Bar-Yosef, and Chris Stringer, 133–51. Cambridge: McDonald Institute for Archaeological Research, 2007.

McBrearty, Sally, and Alison Brooks. "The Revolution That Wasn't: A New Interpretation of the Origin of Modern Human Behavior." *Journal of Human Evolution* 39 (2000): 453–563.

McDougall, Ian, Francis H. Brown, and John G. Fleagle, "Stratigraphic Placement and Age of Modern Humans from Kibish, Ethiopia," *Nature* 433 (2005): 733–36.

McGrew, William C. *The Cultured Chimpanzee: Reflections on Cultural Primatology*. Cambridge: Cambridge University Press, 2004.

———. "Pan Symbolicus: A Cultural Primatologist's Viewpoint." In *Homo Symbolicus: The Dawn of Language, Imagination and Spirituality*, edited by Christopher S. Henshilwood and Francesco d'Errico, 1–12. Amsterdam: John Benjamins, 2011.

Mellars, Paul. "The Character of the Middle-Upper Palaeolithic Transition in South-West France." In *The Explanation of Culture Change: Models in Prehistory*, edited by Colin Renfrew, 255–76. London: Duckworth, 1973.

———. "The Impossible Coincidence: A Single-Species Model for the Origins of Modern Human Behavior in Europe." *Evolutionary Anthropology* 14 (2005): 12–27.

———. "Major Issues in the Emergence of Modern Humans." *Current Anthropology* 30 (1989): 349–85.

Monnier, Gilliane F. "The Lower/Middle Paleolithic Periodization in Western Europe: An Evaluation." *Current Anthropology* 47 (2006): 709–44.

Morin, Eugène, and Véronique Laroulandie. "Presumed Symbolic Use of Diurnal Raptors by Neanderthals." *PLOS ONE* 7 (2012): 1–5.

Peirce, Charles. *The Essential Peirce: Selected Philosophical Writings, Vol. 1*. Edited by E. Houser and C. Kloesel. Bloomington: Indiana University Press, 1992.

Peresani, Marco, Marian Vanhaeren, Ermanno Quaggiotto, Alain Queffelec, and Francesco d'Errico. "An Ochered Fossil Marine Shell From the Mousterian of Fumane Cave, Italy." *PLOS ONE* 8 (2013): 1–15.

Pike, A. W. G., D. L. Hoffmann, M. García-Diez et al. "U-Series Dating of Paleolithic Art in 11 Caves in Spain." *Science* 336 (2012): 1409–13.

Radovčić, Davorka, Ankica Oros Sršen, Jakov Radovčić, and David W. Frayer. "Evidence for Neandertal Jewelry: Modified White-Tailed Eagle Claws at Krapina." *PLOS ONE* 10 (2015): e0119802.

Renfrew, Colin. "The Sapient Behaviour Paradox: How to Test for Potential?" In *Modelling the Early Human Mind*, edited by Paul Mellars and K. Gibson, 11–15. Cambridge: McDonald Institute for Archaeological Research, 1996.

Roebroeks, Wil, Mark J. Siera, Trine Kellberg Nielsena et al. "Use of Red Ochre by Early Neandertals." *Proceedings of the National Academy of Sciences* 109 (2012): 1889–94.

Schwartz, Jeffrey H., and Ian Tattersall. "Fossil Evidence for the Origin of Homo Sapiens." *American Journal of Physical Anthropology* 143.S51 (2010): 94–121.

Shea, John J. "Homo Sapiens Is as Homo Sapiens Was." *Current Anthropology* 52 (2011): 1–35.

Shennan, Stephen. "Population, Culture History, and the Dynamics of Culture Change." *Current Anthropology* 41 (2000): 811–33.

Shennan, Stephen, and J. R. Wilkinson. "Ceramic Style Change and Neutral Evolution: A Case Study from Neolithic Europe." *American Antiquity* 66.4 (2001): 577–93.

Soressi, Marie, Shannon P. McPherron, Michel Lenoire, et al. "Neandertals Made the First Specialized Bone Tools in Europe." *Proceedings of the National Academy of Sciences* 110.35 (2013): 14186–90.

Sterelny, Kim. *The Evolved Apprentice: How Evolution Made Humans Unique*. Cambridge, MA: MIT Press, 2012.

Tobias, P. V. "The Bearing of Fossils and Mitochondrial DNA on the Evolution of Modern Humans, with a Critique of the 'Mitochondrial Eve' Hypothesis." *South African Archaeological Bulletin* 50 (1995): 155–67.

Tomasello, Michael. *A Natural History of Human Thinking*. Cambridge, MA: Harvard University Press, 2014.

Trigger, Bruce. *A History of Archaeological Thought*. Cambridge: Cambridge University Press, 2006.

Trinkaus, Erik, and Pat Shipman. *The Neandertals: Changing the Image of Mankind* (New York: Knopf, 1992).

Val, Aurore. "New Data on the Avifauna from the Middle Stone Age Layers of Sibudu Cave, South Africa: Taphonomic and Palaeoenvironmental Implications." *Quaternary International* 421 (2016): 173–89.

Vandermeersch, Bernard, and María Dolores Garralda. "Neanderthal Geographical and Chronological Variation." In *Continuity and Discontinuity in the Peopling of Europe: One Hundred Fifty Years of Neanderthal Study*, edited by Silvana Condemi and Gerd-Christian Weniger, 113–25. New York: Springer, 2011.

Villa, P., A. Delagnes, and L. Wadley. "A Late Middle Stone Age Artifact Assemblage from Sibudu (KwaZulu-Natal): Comparisons with the European Middle Paleolithic." *Journal of Archaeological Science* 32 (2005): 399–422.

Wadley, Lyn. "Recognizing Complex Cognition through Innovative Technology in Stone Age and Palaeolithic Sites." *Cambridge Archaeological Journal* 23 (2013): 163–83.

White, Leslie. "The Symbol: The Origin and Basis of Human Behavior." *Philosophy of Science* 7 (1940): 451–63.

White, Tim D., Berhane Asfaw, David DeGusta et al. "Pleistocene Homo Sapiens from Middle Awash, Ethiopia." *Nature* 423 (2003): 742–47.

Wolpoff, Milford, Alan Thorne, Jan Jelínek, and Zhang Yinyun. "The Case for Sinking Homo Erectus: 100 Years of Pithecanthropus Is Enough!" In *100 Years of Pithecanthropus: The* Homo Erectus *Problem*, edited by Jens Lorenz Franzen, 341–61. Frankfurt: Courier Forschungsinstitut Senckenberg, 1994.

Wynn, Thomas, and Forrest Tierson. "Regional Comparison of the Shapes of Later Acheulean Handaxes." *American Antiquity* 92 (1990): 73–84.

Zilhão, João, Diego E. Angelucci, Ernestina Badal-Garcíac et al. "Symbolic Use of Marine Shells and Mineral Pigments by Iberian Neandertals." *Proceedings of the National Academy of Sciences* 107 (2010): 1023–28.

THREE

In the Minds of Others

Marcus Baynes-Rock

Here I want to address a question that pertains to a transition in which humans are said to have undergone a paradigm shift in their relations with non-humans: domestication. The question that guides me is this: What can domestication processes tell us about the evolution of wisdom? Of course domestication itself is an ongoing evolutionary process and the inclusion of domesticates in some human societies has had a profound effect on gene expressions in both the non-humans and the humans concerned.[1] For example, in northern European and some sub-Saharan African populations, reliance on milk from domesticates has favored the genetic basis for greater lactase persistence beyond childhood, which makes for a higher percentage of people in these regions who are lactose tolerant.[2] However, for a range of human populations domestication processes have only taken hold in the last fifty to one hundred years and so have had a comparatively limited influence in terms of human evolution. In this case, looking at how domestication influences the evolution of wisdom at a local level would require a very fine-grained analysis beyond what I can achieve here. So instead, I want to look in the other evolutionary direction and consider domestication in its capacity as a product, rather than a producer, of wisdom. This in itself is controversial, considering the deleterious effects domesticatory processes have had on so many humans and animals. But negative consequences notwithstanding, domestication considered as an expression of human wisdom might have a lot to tell us about what it is to be human. While in hindsight it may not have been such a wise move on the part of ancestral humans, the process itself might be rife with expressions of the kind of human wisdom that this book seeks to address. It is these that I want to explore in greater detail.

The kind of wisdom that I am considering has roots that reach deep into the Pleistocene, spanning the millennia over which our human ancestors found a path of divergence from other hominids. In open environments teeming with predators, our bipedal forebears established for themselves a particularly demanding niche. The catchall word "savannah" describes a complex socio-cognitively challenging landscape where predators hunt, scavenge, and defend territories in groups, and have the socio-cognitive skills to match. It is no coincidence that the carnivorous creatures with whom our ancestors have competed over millions of years have a lot of neocortex stuffed into their powerful heads. Lions and hyenas have the highest ratio of neocortex to overall brain volume of all the carnivores, with African hunting dogs coming a creditable fourth.[3] This is because the savannah demands from its social predator-scavengers a suite of characteristics that distinguish them from those solitary types such as leopards who adopt a less confrontational approach to predation. Characteristics such as cooperative problem solving and recognition of third party relationships are not exclusive to apes; in fact, hyenas and dogs outperform chimps in some of these tasks.[4] It was in the savannahs of east Africa that the not-as-wise hominids fell by the wayside, starved or eaten, while those with the cognitive abilities to outwit the lions, hyenas, and dogs became our ancestors.

One of the key characteristics that enabled humans to successfully navigate this challenging ancestral niche is theory of mind.[5] Where humans were forming large groups and competing with large groups of carnivores, theory of mind enabled human ancestors to map representations of other minds onto the world and navigate a path through these complex relationships. Psychologist Simon Baron-Cohen argues that theory of mind was in fact crucial to the emergence of *Homo sapiens*. He lists eight expressions of theory of mind that he argues distinguish humans from other animals.[6] These are intentional communication, repairing failed communication, teaching, intentional persuasion, intentional deception, building shared plans and goals, sharing a focus of attention, and pretending, all of which require an individual to recognize, think about, and try to influence a mental state in another individual. While I would argue, with the support of some very strong evidence, that these are not all exclusive to humans, I would definitely concur that humans have taken these abilities to the next level.[7] As Robin Dunbar illustrates, humans deploy theory of mind to multiple levels of intentionality.[8] Not only can we formulate beliefs about the beliefs of another, but also about another's beliefs about the beliefs of another, and so on up to several orders of intentionality. In this, Dunbar employs an example from Shakespeare's *Othello*, where the great bard intended that his audience realize that Iago knows that Othello believes that Iago is not mistaken when he

Figure 3.1. Spotted hyenas of the research colony at Berkeley cooperating to pull on ropes which together unlatch a tilting platform which delivers a food reward. Photograph reproduced with permission from Christine Drea.

says Desdemona desires Cassio. Baron-Cohen goes as far as to argue that theory of mind is a necessary condition for symbolic language.[9] This is because without being able to apprehend that another has a particular belief, it is impossible to try to influence that belief with any sort of symbolic communication. The evidence for this is found in Autism Spectrum Disorder (ASD), which in varying degrees inhibits a person's ability to understand that another has particular mental states. Autism is also associated with language difficulties.[10]

Autism spectrum disorder is very telling in terms of the evolution of theory of mind. This is because it stems from a very recently evolved DNA sequence associated with language and cognition, which is claimed to be specific to *Homo sapiens*.[11] It is ironic that the little collection of amino acids that is a key element of what I am presenting here as an expression of human wisdom is also, by its recently evolved nature, integral to human wisdom's malfunctioning.[12] In this way we can think of theory of mind as it is manifested in *Homo sapiens* as a recently developed racing engine. When it works, it does some amazing things and is a wonder to behold. But, being so innovative and finely tuned, it is highly prone to malfunctioning. Thus, we could say that autism is symptomatic of a pushing of boundaries in terms of the limits to which humans can take cognition. And as I will demonstrate, these limits are constantly being challenged.

In evolutionary terms, there is no point in asking when full blown theory of mind, and by extension language, emerged. It was and is a process with no particular beginning and no foreseeable end. All we can say is that sometime after the separation of the human and chimp lineages, humans thought more in terms of an environment populated with other minds, both human and non-human. While their primate cousins of the forest could only fear the darkness and its unknown dangers, ancestral humans learned not only to imagine the kinds of creatures lurking in the darkness but also what those creatures' thoughts and intentions might be. Moreover, they came to apply the capacity for "minding" in multiple interactions with their surrounding world, transcending modes of associative learning to the point where the mental states of others guided their logics.[13] Our ancestors' ability to apply mindedness to predators, prey, forests, and phenomena enabled them to navigate a complex niche of opportunities and challenges which extended human sociality beyond the group huddled around the campfire to a wide world of creatures and things, seen and unseen. Of this we can be certain because ethnographies of contemporary hunter-gatherers consistently find that they live in social worlds populated with all manner of minded entities from lizards to predators to weather events. By this I am not saying that contemporary hunter-gatherers are emblematic of our ancestral past. Rather, these people inherited from our shared ancestors the capacity to apply theory of mind to the non-human world and, due to their environmental circumstances, persist in its application.

From his ethnographic work among Yukagir hunters of Siberia, Rane Willerslev describes animist worlds of hunters and hunted where minds are so categorically unstable as to cross species.[14] In order to hunt successfully, hunters must enter the minds of their prey. One hunter named Taishin explains to Willerslev the key to successfully trapping sable. "To place your traps well, you must be curious like the sable. You need to think like this: 'What would attract my curiosity as a sable?' This is how you need to think."[15] The same applies to hunting elk. For the Yukaghir, elk are precarious beings liable to run away at the slightest mistake by a hunter. However, the kinds of mistakes possible include far more than making a noise and startling the animal. Prior to going out to hunt elk, Yukaghir hunters speak about their intentions with euphemisms. They say, "I'm going for a walk" or "Let's go and look at the 'big one.'"[16] They do this in case an elk overhears them speaking. "They [the elk] talk with one another. If one of them has heard what you said, it will tell the rest and they will all move away."[17] Thus, the hunters constantly strive to anticipate the elk's mental processes in order to hunt successfully, but this in turn is a cognitively precarious business. They believe that the process of mental projection comes with a risk of being transformed into that which they seek to understand. Hunters relate experiences of being incorporated into elk societies or of unselfconscious over-hunting to the point of realization that they are turning into wolves.[18] Indeed it is so easy for them to project their minds into the worlds of animals that the rest of their bodies are dangerously liable to follow. Willerslev describes this as a kind of mental balancing act. The hunter must find within himself the perspective of an elk while at the same time being aware of his elk-perspective so that he can set it aside in an instant and shoot his quarry. His success depends on what Willerslev calls his ability to hold a "double perspective," something we can also call a reflexive theory of mind, or theory of mind to a second order of intentionality directed at the self.[19]

Examples of this kind of minding of the non-human world abound in anthropological literature. While living among the Avila Runa of Upper Amazonia, Eduardo Kohn was given advice on which sleeping position to adopt. "Sleep face up! If a jaguar comes, he'll see you can look back at him and he won't bother you."[20] Here the Runa enter into the jaguars' mental states and conceive of the different ways in which jaguars might represent humans based on something akin to Martin Buber's subject/object distinction.[21] While Buber might hold that animals are unable to fully apprehend the subjectivity of another, Runa have no such inhibition. Moreover, the mentalizing activities of the Runa, and many Amazonians for that matter, extend to complex accounts of other animals' phenomenal worlds. Within these accounts, different creatures' perspectives are aligned with an ultimate reality, but each is manifested differently depending on species. This opens up a vast "ecology of selves" in which their perspectives on the minds of monkeys, peccaries, and leafcutter ants guide people's pragmatic ways of engaging with the world, which in turn enable them to "harness some of its plenty."[22] The Runa are well aware of the efficacy of their mentalizing. They complain of hunters suffering from *ah^{h}huas* or soul-blindness, a condition usually found in men with pregnant wives, which prevents hunters' adopting the perspectives of their prey. In fact, so debilitating is this condition that the prey refuse to appear before soul-blind hunters' guns or enter into their traps.[23] In a classic train of intentionality, of which Shakespeare would be proud, Kohn describes how the people believe that the animals know that these hunters no longer perceive the perspectives of their prey.

Another example of a people being guided in their practical affairs by complex perspectives on other minds comes from the circum-polar region of North America. Here Richard Nelson describes the more-than-human social worlds of the Koyukon. These people not only take precautions based on the minds of animals but also on those of the animals' spirit masters. Not only do Koyukon deploy theory of mind to anticipate the mental processes of the animals they intend to hunt, but after a successful hunt these people take care not to offend the animals' spirit masters.[24] For the more spiritually powerful species such as wolverine, this involves giving the dead animals elaborate funerals, dressing them up, offering food, and greeting them as honored guests. People are very careful to maintain goodwill with animals' spirit masters because without such amicable relations the animals themselves might withdraw their cooperation in the hunt.[25] For a subsistence hunter this would be a disaster. The fact that these people and their perspectives persist in a challenging environment is testament to the ways in which theory of mind as an expression of human wisdom is integral to human flourishing.

In a devastating critique, anthropologist Tim Ingold has pulled these hunter/gatherer ontologies out from under the more traditional anthropological conceptions based on Western dualisms.[26] In cases where hunter/gatherers describe their engagements with non-human animals to anthropologists, they have not traditionally been taken at face value. All this talk of hunters being seduced by their prey and particular animals withholding hunting privileges is so at odds with the nature/culture dualisms of Western thought that it can only be explained away as metaphorical.[27] Those grudge-bearing wolverines and alluring elk must surely be symbolic constructions enacted in the human realm to explain away phenomena in a separate "natural" realm, and only anthropologists are able to connect the disparate pieces. But as Ingold demonstrates, this metaphor-based explanation is itself based on Western dualisms to which the hunter gatherers do not themselves subscribe. For them, there is no separation of nature and culture—there is just a world of humans and other beings with whom they are socially engaged. They certainly maintain a categorical distinction between humans and other animals, and in fact are very focused on the distinction, but it is one of bodily manifestation and perception rather than the acute natural/cultural distinction in the Western tradition which itself is the bedfellow of mind/body dualisms.[28] When hunter-gatherers talk of their engagements with humans and other animals, there is indeed a distinction between the kind of minds with which they are engaging, but not in terms of the realm within which they are engaged. Both are social. Ingold concludes: "there can, then, be no break between social and ecological relations; rather, the former constitute a subset of the latter."[29] This is because humans not only think about the minds of other animals; it is that they think themselves "into" the minds of other animals. Here Agustín Fuentes challenges Thomas Nagel's contention, based on the different perceptual worlds of bats and humans, that we cannot know what it is like to be another species. On the one hand, Fuentes argues that we can know what it is like to be apes because we are apes.[30] On the other, the accuracy of our cross-species mental attributions is not as crucial as whether they facilitate our engagements with those others in the lived environment.

In terms of the ways in which humans conceive of animals, Ingold describes a profound transformation that occurred with the shift from hunter/gatherer lifeways to those of animal domestication. Whereas Western conceptions of domestication see it as a process whereby nature is appropriated and brought into the human realm, Ingold instead conceives of it as a shift in human relations with the environment in which they are immersed. For hunter-gatherers, animals are precarious beings. They and their spirit masters, or essences, must be placated in order that the animals present themselves in future hunting forays. These are social relationships between hunters and entities in the environment based on trust and reciprocity, the same sorts of relations hunters have with their fellow humans.[31] But in the process of domestication, the animals are denied their autonomy. Ingold argues that the trust between hunter and animal is thrown out in favor of physical means by which the animals are brought under human control. While both "economies" involve an immersion in the world of other beings and a recognition of the animals' autonomy, domestication brings a radical shift by which reciprocity no longer figures in the relationship.[32]

However, there are ethnographic examples that show Ingold's thesis could use some fine tuning, with particular reference to mind-reading. Anthropologist Natasha Fijn, who conducted fieldwork among herders in Mongolia, argues that trust and reciprocity are crucial to herders' relations with their animals. Here it is wolves that mediate the relationship. The herders see themselves as providing protection for their herd animals from wolves and other threats, and they spend a lot of time and energy achieving this.[33] The animals reciprocate by providing milk and eventually meat. There are no fences or physical means of control—it is through understanding the needs and intentions of their animals and establishing trust that the herders maintain the goodwill of their livestock. The herders see this relationship as reciprocal.[34] Of course the killing of the animal is the end point of the relationship, but this in fact creates a closer correspondence with the relations that Ingold confines to the realm of the hunter-gatherer. In the same way that Koyukon hunters propitiate their prey, Mongolian herders exchange care and protection for the presence of the herd animal at the time when they need to kill it.

The findings of Fijn have parallels with the results of my own fieldwork among the Oromo of Ethiopia. There, as with Mongolia, there are no fences to keep the animals contained. The horses, cattle, donkeys, sheep, and goats are free to wander as they please during the day, although the more valuable animals are accompanied by a boy to protect them from thieves. But regardless of whether they have a human chaperone, these animals dutifully return to the house at the end of the day and wait to be let inside. Here it is a combination of theory of mind and spotted hyenas that binds the relationship between livestock animals and farmers. The people know that their animals know that there are hyenas lurking in the landscape and so they do not need to build fences. They simply need to provide some form of protection during the night when hyenas are about, and with the setting of the sun the animals are lining up waiting to be let in. Horses can be sent out to run around with other people's horses, donkeys can be trusted to find their way home with a load of goods from the market, and goats and sheep can be relied upon to remain close to the house. This is the combination of "autonomy and dependency" that according to Ingold characterizes trust.[35] It abounds among Oromo herders because they apply their cognitive capacities to mediating relations between domesticated animals and predators. Nevertheless we should not throw Ingold's baby out with the bath water. While people do trust their animals to reciprocate in their shared project of mutual flourishing, they still exercise a form of control—only it is not an overtly physical means of control such as a fence or rope. Instead it is a kind of mental manipulation. Oromo people (to varying degrees) are attuned to the thoughts and habits of their animals in relation to their phenomenal worlds, and they capitalize on this. Through application of theory of mind they navigate a social world in which they establish a mutual dependence with their livestock in order that they and their animals might thrive. Given this, the domination that Ingold argues emerged with animal domestication might be re-framed as a more effective, up-close form of manipulation than that of the hunter/gatherer. They are not so different after all.

So we can see how animal domestication bears the marks of human wisdom in that it is wound up with theory of mind and its application across species boundaries. However, domestication comes in all shapes and sizes and there are various levels of trust invested in domesticated animals. For example, in my home region in northern New South Wales, the farming paradigm differs from that of Ethiopia. Here cattle farmers employ a technological strategy in order that their animals stay around: barbed wire. This kind of fencing is abundant in Australia, so much so that you cannot walk thirty paces from a road without getting snagged on a fence. Barbed wire symbolizes everything about colonial, settled farming. It speaks of control of land—there are no rights of way here; it speaks of economies of scale—there are no children guiding mixed herds,

Figure 3.2. Livestock animals grazing in a communal pasture in Oromia. Photograph by Marcus Baynes-Rock.

but single breeds in their hundreds, wearing numbered ear tags, and all contained within prickly edged quadrangles. In fact, the bigger the plot of land the cheaper the fencing per hectare, and economy is the bottom line. There is no trust here. It is inconceivable to the farmers that the animals would show any site fidelity if the fencing was removed. Cattle are units of currency that threaten to aimlessly wander off and the only way to control them is to fence them in. People can be trusted to the end of the earth—at least the neighbors can, not the government and the "greenies." But cattle are mindless wanderers, vacuously chewing their cud and following the pickup from paddock to paddock, hooked on the scent of a bale of hay. Or are they?

Given the supposed lack of trust that Australian farmers have in their livestock animals, I wanted to find out exactly what, if anything, they thought was going on in those bovine/ovine/equine heads. I predicted that the modern farming paradigm with its fencing, ear tags, breeding programs and economies of scale would be, in terms of social relations with animals, further removed from Oromo and Mongolian pastoralists than those forms of domestication are from hunter/gatherers. However, when I began talking to some farmers I found that they readily applied mental states to their animals and again it was predators that mediated mentalizing of their livestock animals. In discussions about the problems that wild dogs posed to their cattle, horses, and sheep, farmers often used theory of mind as a means to understanding their animals' modes of dealing with the threat of predation. For example, one farmer, in discussing how wild dogs single out calves, said, "If the calf is injured and the cow comes back, she sees it's no good and leaves it." Here it is irrelevant whether the farmer is describing an instinctive reaction of the cow to the sight of her mortally injured calf or a decision making process on the part of the cow. What is crucial here is that he is describing a mental process of perception and action. But it does not end with attributions of decision making processes; there is also much talk of how these are influenced by emotional states and states of knowledge:

> They get particularly . . . the cows if they've got young calves, they get anxious. Very you know . . . They're more alert and looking around and if we have one of our dogs with us, we're checking them and really they're very wary of the dogs.

This theme emerged with many of the farmers interviewed where they described their livestock animals as acting on states of anxiety in response to dogs.

> The other interesting thing is that over the last two months the sheep have changed their behaviour—the sheep no longer go up to where they were going. They're not going so far—they're sticking close by. So the whole behaviour—the animals' behaviour has changed. We think that's because of the dogs because they (the sheep) want to be close to their security.
>
> We've got some paddocks way over the back and if there's strife over there we'd have no idea. So it's better if our cows are very wary and have that fear of dogs. So when they do see dogs they can go "Right, this is unusual, this isn't meant to be happening. We'd better stand up for ourselves and do something."

Of course, the farmer quoted here does not believe that cows talk among themselves; instead, this kind of anthropomorphizing is aimed at giving me, the interviewer, an understanding of the mental processes underlying the behaviors. Whether or not this is an accurate summation of the cows' mental processes is not central to my thesis. What is interesting in these data is that Australian farmers operating within a paradigm that favors objectification still resort to the application of mental states to their livestock animals. They think in terms of how their cattle and sheep are thinking and how these thoughts guide action, suggesting that they are engaged in much more than owner/object relations with their animals. Indeed, even a cattle farmer from the northern tablelands with 2,400 acres and 1,200 head of cattle had no compunctions describing how his animals reacted emotionally to dogs.

> They'll start bellowing and you'll see the cattle will. Once cows start calving they'll get agitated and they'll run around and they'll be . . . they'll group together and you can see. You know you can see . . . you

> can see . . . where . . . if you go out and the fence might be pushed or something and you know you sort of saying why did they all rush to the fence? So something's obviously spooked them.[36]

Of course this application of mental states to livestock animals might have a practical aspect. After all, if a farmer can better understand the mental states of her cattle or sheep in response to the presence of wild dogs, then this might foster better herd management strategies, although this is yet to be tested. Nevertheless, whether or not this mentalizing of herds has any practical value, my argument here is that these farmers would engage in it regardless. This is because the application of theory of mind to the non-human world is so integral to human wisdom that it is inveterate. Regardless of the cultural context, humans will find minds in all manner of entities from dogs, to cows, to tractors, to natural phenomena. This is compulsive theory of mind, and it is a key aspect of the human condition.

THE SHAPES OF HUMAN WISDOM

In 1944, psychologists Fritz Heider and Marianne Simmel of Smith College published the results of an experiment that they created to explore the ways in which humans perceived others' behaviors and personal qualities.[37] What was different about their experiment was that instead of using human faces or gestures as target stimuli, they used shapes. Heider and Simmel created a now-famous animated sequence of one-and-a-half minutes in which two triangles and a circle are depicted moving around a field containing a rectangle.[38] The animators created a kind of narrative in which the shapes interact and exhibit movements which betray—to most humans at least—goal directedness and affective states. This they did successfully because all but one of the seventy subjects in their experiment described the sequence in terms of the shapes being intentional persons engaged in a social interaction.[39] By way of example they provide one person's account of the animation:

> Triangle number-one shuts his door (or should we say line) and the two innocent young things walk in. Lovers in the two-dimensional world, no doubt; little triangle number-two and sweet circle. Triangle-one (here-after known as the villain) spies the young love. Ah! . . . He opens his door, walks out to see our hero and his sweet. But our hero does not like the interruption (we regret that our actual knowledge of what went on at this particular moment is slightly hazy, I believe we didn't get the exact conversation), he attacks triangle-one rather vigorously (maybe the big bully said some bad word).[40]

The above description is in fact atypical in that it holds that the shapes are shapes. In other accounts, the triangles and circle are unashamedly described as man, woman, husband, wife, child, bird, and, in the version

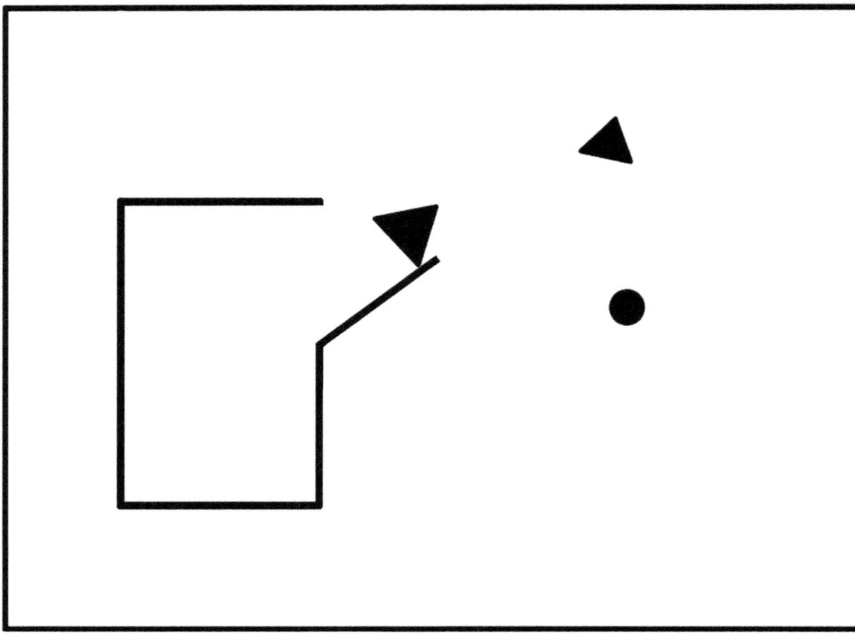

Figure 3.3. Still image from Heider and Simmel's animation. Redrawn by Marcus Baynes-Rock from Heider and Simmel 1944.

of the experiment where the animation was played backward, Evil. What is even more interesting is not just that so many people described the shapes in terms of intentional persons, and their movements in terms of intentions and emotions, but that their accounts were remarkably convergent in terms of the intentions and personalities that were ascribed to the shapes. This was especially the case with the large triangle who is described by 97 percent of the subjects as either "Aggressive, warlike, belligerent, pugnacious, quarrelsome, troublesome, mean, angry, bad-tempered, temperamental, irritable, quick to take offense, bully, villain, taking advantage of his size, picking on smaller people, dominating, power-loving, or possessive."[41]

This is a remarkable degree of consensus, and should lead one to believe that there must be something "right" about the subjects' interpretations. However, fortunately for my argument, there was one subject in Heider and Simmel's experiment whose description of the animation differed significantly from the other people's characterizations:

> A large solid triangle is shown entering a rectangle. It enters and comes out of this rectangle, and each time the corner and one-half of one of the sides of the rectangle form an opening. Then another, smaller triangle and a circle appear on the scene. The circle enters the rectangle

> while the larger triangle is within. The two move about in circular motion and then the circle goes out of the opening and joins the smaller triangle which has been moving around outside the rectangle. Then the smaller triangle and the circle move about together and when the larger triangle comes out of the rectangle and approaches them, they move rapidly in a circle around the rectangle and disappear. The larger triangle, now alone, moves about the opening of the rectangle and finally goes through the opening to the inside. He (sic!) moves rapidly within, and, finding no opening, breaks through the sides and disappears.[42]

This entirely accurate account is so at odds with the other sixty-nine accounts that it is tempting to suggest that it is wrong in some respect. But how can it be wrong when it so accurately describes what is in the film? I suggest that it is wrong only insofar as it represents a failure to engage with the mental states that the animators imbibed into the shapes. Heider and Simmel animated the shapes in a particular way with the assumption that people would ascribe mindedness and personhood to non-human entities bearing goal-directed movements. The above, rather dry account represents a failure to deploy theory of mind beyond the human and by extension a failure to engage with a social interaction that is apparently obvious to the majority of humans. In light of later experiments, it seems extremely likely that the subject who gave the above account had some degree of autism.

Considering the way in which the Heider and Simmel animation so easily elicits theory of mind in normal subjects, it is unsurprising that it has been used in experiments on people with autism. The utility of the animation, they argue, is that it controls for some factors that confound other theory of mind tasks for people with autism. For example, in the Sally and Ann task, which is designed to test knowledge of false belief, cognitively able people with autism prove to be as adept at the task as non-clinical subjects, yet they still seem to fail in real life social situations.[43] This disparity in success, some researchers believe, could be a result of the application of problem solving skills rather than theory of mind. Thus, because the animation task is aimed at eliciting attributions of mental states and personalities which cannot be arrived at simply through problem solving skills, it better assesses the degrees to which people with autism are able to employ theory of mind.[44] After controlling for language difficulties and so forth, these experiments show that people with autism simply do not ascribe the same level of intentionality, affect, and personhood to the geometric shapes, irrespective of IQ and metalinguistic skills.[45] In this example from an experiment by Ami Klim of the Yale Child Study Centre, an autistic child's account of the Heider and Simmel animation parallels that of the outlier in the original 1944 experiment:

> The big triangle went into the rectangle. There were a small triangle and a circle. The big triangle went out. The shapes bounce off each other. The small circle went inside the rectangle. The big triangle was in the box with the circle. The small triangle and the circle went around each other a few times. They were kind of oscillating around each other, maybe because of a magnetic field. After that, they go off the screen. The big triangle turned like a star—like a Star of David—and broke the rectangle.[46]

Even though this child had a verbal IQ that was comparable to non-clinical children, his account is markedly different from the accounts of those children. There is no indication of an ascription of mental states. In circumstances where non-clinical children use terms such as shy, scared, jealous, and upset to account for the movements of the shapes, autistic children simply do not see these states of being in the animation. For non-clinical subjects, the triangles and circle are qualitatively different from the rectangle: they are subjects and object; they act upon each other and upon the rectangle, while the rectangle remains an object. For people with autism, the other shapes are only different from the rectangle insofar as they exhibit movement.

In another experiment using the same animation, Dermot Bowler and Evelyn Thommen sought the finer-grained differences between children with ASD and non-clinical children in attribution of mental states. While some autistic children in their sample capably applied mental state terms to the shapes, these children failed to recognize any underlying goal-directedness in the animated triangles and circle.[47] It was as if their schooling had taught them to apply mental state terminology, but they had no internal compass guiding them as to which states fitted which terms. That both autistic and non-clinical children exhibited a floor effect in this experiment suggested that the Heider and Simmel animation was not ideal for these kinds of experiments.[48] This led Abell and colleagues to construct a new experiment, again with animated shapes but more nuanced.[49] This time the subjects were shown three animations depicting variously random movements, synchronized movements, and intentional movements of two triangles. In this way they hoped to be able to separate the application of mental states to any moving objects from application of mental states to (apparently) intentional objects. In the results they found that autistic children who passed first and second order false belief tasks performed poorly on the theory of mind tasks using animations. Even where they used mentalizing terms, these were applied inappropriately and in some case even to the animation of triangles moving randomly.[50] These researchers suggest that people with high functioning autism have the neural architecture to be able to use mental state terminology but lack the neural connections to be able to apply it under the right circumstances. The underlying etiology for this, they argue, is the "prototype" nature of the neural networks involved in theory of mind that I referred to earlier.

The current research on autism is very telling in terms of the evolution of wisdom. It tells us that theory of mind—the form of wisdom that humans evolved in order to navigate the complex social networks among humans and non-humans in open environments—is a fragile thing. But it is also something that is immensely powerful in that it takes hold of people's imaginations and finds for them all manner of minds in their phenomenal worlds. This is what emerges out of studies of domestication processes where theory of mind applied to the non-human world is not only integral to these processes in many parts of the world but inveterate; even where the utility of theory of mind is limited, people cannot help but ascribe mental states to the animals that they otherwise control. This inveterate theory of mind is made most explicit when we find that people ascribe mental states and personalities to such things as animated triangles. To say that a two-dimensional triangle moving about a screen is angry or jealous is a patently human thing to do, but at the same time it is objectively wrong. The triangle might move as an angry human would but it is in no way angry. And yet this application of a mental state and even a personality to a triangle is considered correct and appropriate. Why is this so? It is because it is a functional interpretation in terms of navigating the human social world. It serves a human to ascribe intentionality to such things because this is where theory of mind has taken us, and autism tells us not only how novel and fragile are the neural substrates that underlie this excess of theory of mind but also how it is less an access to reality than it is a window into a social world of humans, non-humans, and natural phenomena. This brings me to the next interesting aspect of theory of mind: people with autism are less likely to believe in God.

Psychologist Ara Norenzayan set out to test the hypothesis that the cognitive foundations underlying belief in God and other supernatural entities lay in ordinary human social cognition: specifically, mentalizing. This hypothesis held that theory of mind was a necessary component—although not a sufficient cause—of belief in God.[51] Considering that people with autism exhibited deficiencies in application of theory of mind, Norenzayan decided to test whether autistic people held the same levels of belief in God as non-clinical people. This was done in a series of experiments using scales designed to elicit measures of religiosity and belief in a personal God. After controlling for such things as lower levels of religious attendance (in autistic subjects), religious involvement causing religious beliefs, scientific interest, agreeableness, and conscientiousness, the study of Canadian and American subjects found that the hypothesis was supported—that there was an inverse link between autism and belief in God, and furthermore that this was down to the autistic subjects' lack of capacity for mentalizing.[52] Norenyayan went on to draw correspondences between the proven (although theoretically debated) gender gap in belief in God and the close association of the deficiency in empathy in males and mentalizing deficits in people with autism, the majority of whom are also male. What these findings do is to take evolved wisdom as expressed in theory of mind as an adaptive way of thinking outside the human. It weaves cognition into a complex web of relations whereby theory of mind becomes integral not only to the cohesion and complexity of human social groups but as a mediator in human relations with the non-human world. This is not just the case with non-human animals in human environments and through domestication, but also in what Jesse Bering calls "existential theory of mind."[53] Theory of mind fostered relations beyond humans and animals to the inanimate and the abstract, whereby this recently evolved, delicate, and novel mode of cognition enabled humans to establish a relationship with God.

So what does the domestication of animals tell us about the evolution of wisdom? On the one hand, it shows us how theory of mind is so ingrained in most humans that regardless of its efficacy, it finds expression in their relations with domesticates. But was theory of mind, as an expression of human wisdom, crucial to animal domestication? Here I suggest that we should not just privilege theory of mind with respect to our relations with the world, but should consider autism as more than a simple pathology born of a novel and sensitive apparatus in normal humans. Anthropologist Penny Spikins suggests that autism, a highly heritable condition, has itself been crucial to the evolution of wisdom, albeit a technical, practical form of wisdom.[54] Citing a list of well-known innovators, all of whom exhibited features typical of autism or Asperger's, Spikins argues that this condition has fostered the flourishing of human groups. Pointing to cave art and micro blade technologies of the late Pleistocene, she suggests that the kind of focus, attention to detail, and capacity for innovation that lay behind these modes of expression were very likely capacities of people with some measure of autism. This is what led Fitzgerald to say of these savants that "the human race would still be sitting around in caves chattering to each other if it were not for them."[55] From this perspective, theory of mind was a crucial component in human evolution, but it was also its malfunctioning that fostered the emergence of *Homo sapiens* as the dominant species on Earth. Thus I suggest that autism might be regarded as a contribution to the diversity of humans which in itself is adaptive. And it even has a place in domestication processes. In the list of famous autistic people that Spikins cites is Temple Grandin, an award-winning scientist whose ability to understand cattle better than regular people has had profound effects on livestock husbandry across the world.[56] This suggests that the evolved form of wisdom I am considering here, as expressed in theory of mind across species, is elicited in domestication processes but is not crucial to them, because as Grandin demonstrates with such salience, other forms of reasoning can be more effective. Just because most humans are inveterate mentalizers and able to understand this capacity in each other, this does not mean that our mentalizing is superior in terms of navigating more than human social worlds.

NOTES

1. Richard Francis, *Domesticated: Evolution in a Man-Made World* (New York: W. W. Norton, 2015), 310–11.

2. Dallas M. Swallow, "Genetics of Lactase Persistence and Lactose Intolerance," *Annual Review of Genetics* 37 (2003): 197–219; Sarah A. Tishkoff, Floyd A. Reed, Alessia Ranciaro, Benjamin F. Voight, Courtney C. Babbitt, Jesse S. Silverman, Kweli Powell et al., "Convergent Adaptation of Human Lactase Persistence in Africa and Europe," *Nature Genetics* 39.1 (2007): 31–40.

3. Robin I. M. Dunbar and Julian Bever, "Neocortex Size Predicts Group Size in Carnivores and Some Insectivores," *Ethology* 104 (1998): 698.

4. See, for example, Christine M. Drea and Alissa N. Carter, "Cooperative Problem Solving in a Social Carnivore," *Animal Behaviour* 78 (2009): 967–77; Anne L. Engh et al., "Patterns of Alliance Formation and Postconflict Aggression Indicate Spotted Hyaenas Recognize Third-Party Relationships," *Animal Behaviour* 69 (2005): 209–17.

5. Here I consider theory of mind as complementary to joint attention; Tomasello holds these to be "related but distinct steps" in terms of human socio-cognitive development. The distinction rests on theory of mind being the perception that others have particular beliefs while joint attention is the perception that others have particular intentions. See Michael Tomasello, "Joint Attention as Social Cognition," in *Joint Attention: Its Origins and Role in Development*, ed. Chris Moore and Philip J. Dunham (New York: Psychology Press, 2014), 104.

6. Simon Baron-Cohen, "Evolution of a Theory of Mind?" in *The Descent of Mind: Psychological Perspectives on Hominid Evolution*, ed. Michael C. Corballis and Stephen E. G. Lea (Oxford: Oxford University Press, 1999), 265–72.

7. See Marc Bekoff and Colin Allen, "Intentional Communication and Social Play: How and Why Animals Negotiate and Agree to Play," in *Animal Play: Evolutionary, Comparative, and Ecological Perspectives*, ed. Marc Bekoff and John A. Byers (Cambridge: Cambridge University Press, 1997); Robert W. Mitchell, *Pretending and Imagination in Animals and Children* (Cambridge: Cambridge University Press, 2002).

8. Robin Dunbar, *The Human Story: A New History of Mankind's Evolution* (London: Faber and Faber, 2004), 121.

9. Baron-Cohen, "Evolution of a Theory of Mind?" 266.

10. Antonio Benítez-Burraco, Wanda Lattanzi, and Elliot Murphy, "Language Impairments in ASD Resulting from a Failed Domestication of the Human Brain," *Frontiers in Neuroscience* 10 (2016): 373.

11. See Xander Nuttle et al., "Emergence of a *Homo Sapiens*-Specific Gene Family and Chromosome 16p11.2 CNV Susceptibility," *Nature* 536 (2016): 205–29.

12. Elliot Murphy and Antonio Benítez-Burraco, "Language Deficits in Schizophrenia and Autism as Related Oscillatory Connectomopathies: An Evolutionary Account," *Neuroscience and Biobehavioral Reviews* (in press): 3; Sara Reardon, "Inside the Brain of a Neanderthal," *New Scientist* 219.2926 (2013): 6–7; Roberto Toro et al., "Key Role for Gene Dosage and Synaptic Homeostasis in Autism Spectrum Disorders," *Trends in Genetics* 26 (2010): 363–72.

13. This is not to suggest that other animals are solely guided by associative learning mechanisms as there is a considerable body of evidence—albeit not unambiguous—to demonstrate complex cognitive functions in non-humans, including inferential reasoning, sensitivity to ambiguity, and metacognition (see Mark Haselgrove, "Overcoming Associative Learning," *Journal of Comparative Psychology* 130.3 (2016):

226–40). What I am suggesting is that human ancestors became more reliant on inferences of the minds of other animals by analogy to their own minds than by learned associations through experience of animal behaviors.

14. Rane Willerslev, *Soul Hunters: Hunting, Animism, and Personhood among the Siberian Yukaghirs* (Berkeley: University of California Press, 2007).

15. Ibid., 91.

16. Ibid., 100.

17. Ibid., 75.

18. Ibid., 90.

19. Ibid., 97.

20. Eduardo Kohn, *How Forests Think: Toward an Anthropology Beyond the Human* (Berkeley: University of California Press, 2013), 1.

21. Martin Buber, *I and Thou* (Edinburgh: T & T Clark, 1970 [1937]), 112.

22. Kohn, *How Forests Think*, 17, 95.

23. Ibid., 123.

24. Richard K. Nelson, *Make Prayers to the Raven: A Koyukon View of the Northern Forest* (Chicago: University of Chicago Press, 1982), 149.

25. Ibid., 153.

26. Tim Ingold, *The Perception of the Environment: Essays on Livelihood, Dwelling, and Skill* (London: Routledge, 2000).

27. Ibid., 44.

28. Ibid., 41.

29. Ibid., 60.

30. Agustín Fuentes, "The Humanity of Animals and the Animality of Humans: A View from Biological Anthropology Inspired by J. M. Coetzee's Elizabeth Costello," *American Anthropologist* 108.1 (2006): 124–32.

31. Ingold, *Perception of the Environment*, 72.

32. Ibid., 75.

33. Natasha Fijn, *Living with Herds: Human-Animal Coexistence in Mongolia* (Cambridge: Cambridge University Press, 2011), 210.

34. Ibid., 241.

35. Ingold, *Perception of the Environment*, 69.

36. Transcript from an interview that took place in April 2016 in the New England Tablelands, New South Wales.

37. Fritz Heider and Marianne Simmel, "An Experimental Study of Apparent Behavior," *The American Journal of Psychology* 57.2 (1944): 243–59.

38. For an example of the animation, see Julie Freeman, "Heider and Simmel Animation," *Vimeo* video, 1:12, posted by "Translating Nature," September 5, 2012, https://vimeo.com/48908599. In one iteration of this animation, a group of comedians give a commentary on their perceptions of the actions of the shapes. See "Out of Shape," *YouTube* video, 4:25, posted by University of Southern California Institute for Creative Technologies (USCICT), February 12, 2014, https://www.youtube.com/watch?v=ZAnt9II-5Co.

39. Even when the film was played backward all but two of the subjects gave accounts of the shapes as minded actors.

40. Heider and Simmel, "An Experimental Study," 247.

41. Ibid., 248.

42. Ibid., 246.

43. The Sally and Anne task tests the ability of people to attribute a state of false beliefs to a fictional character: Sally. Initially, Simon Baron-Cohen, Alan Leslie, and Uta Frith found that sixteen out of twenty children with autism failed the belief task, suggesting that people with autism fail to employ a theory of mind. However, Ami Klin points to series of studies where people with higher functioning autism and Asperger's succeed on theory of mind tasks at different levels of complexity but still fail in real-life social situations.

44. Ami Klin, "Attributing Social Meaning to Ambiguous Visual Stimuli in Higher-Functioning Autism and Asperger Syndrome: The Social Attribution Task," *Journal of Child Psychology and Psychiatry* 41.7 (2000): 831–46.

45. Dermot M. Bowler and Evelyne Thommen, "Attribution of Mechanical and Social Causality to Animated Displays by Children with Autism," *Autism* 4.2 (2000): 147–71; Klin, "Attributing Social Meaning."

46. Klin, "Attributing Social Meaning," 840.

47. Bowler and Thommen, "Attribution of Mechanical and Social Causality," 161.

48. Ibid., 161.

49. Frances Abell, Francesca Happé, and Uta Frith, "Do Triangles Play Tricks? Attribution of Mental States to Animated Shapes in Normal and Abnormal Development," *Cognitive Development* 15.1 (2000): 1–16. See also Fulvia Castelli, Chris Frith, Francesca Happé, and Uta Frith, "Autism, Asperger Syndrome and Brain Mechanisms for the Attribution of Mental States to Animated Shapes," *Brain* 125 (2002): 1839–49.

50. Abell, Happé, and Frith, "Do Triangles Play Tricks?" 9.

51. Ara Norenzayan, Will M. Gervais, and Kali H. Trzesniewski, "Mentalizing Deficits Constrain Belief in a Personal God," *PLoS One* 7.5 (2012): 1.

52. Norenzayan, Gervais, and Trzesniewski, "Mentalizing Deficits," 4.

53. Jesse M. Bering, "The Existential Theory of Mind," *Review of General Psychology* 6.1 (2002): 3–24.

54. Penny Spikins, "Autism, the Integrations of 'Difference' and the Origins of Modern Human Behaviour," *Cambridge Archaeological Journal* 19.2 (2009): 179–201.

55. Quoted in J. Griffin, "Autistic Tendencies: The Consequences for Our Culture," *Human Givens* 13.4 (2006): 17.

56. Spikins, "Autism," 11.

BIBLIOGRAPHY

Abell, Frances, Francesca Happé, and Uta Frith. "Do Triangles Play Tricks? Attribution of Mental States to Animated Shapes in Normal and Abnormal Development." *Cognitive Development* 15.1 (2000): 1–16.

Baron-Cohen, Simon, Alan M. Leslie, and Uta Frith. "Does the Autistic Child Have a 'Theory of Mind'?" *Cognition* 21 (1985): 37–46.

———. "Evolution of a Theory of Mind?" In *The Descent of Mind: Psychological Perspectives on Hominid Evolution*, edited by Michael C. Corballis and Stephen E. G. Lea, 261–77. Oxford: Oxford University Press, 1999.

Bekoff, Marc, and Colin Allen. "Intentional Communication and Social Play: How and Why Animals Negotiate and Agree to Play." In *Animal Play: Evolutionary, Comparative, and Ecological Perspectives*, edited by Marc Bekoff and John A. Byers, 97–114. Cambridge: Cambridge University Press, 1997.

Benítez-Burraco, Antonio, Wanda Lattanzi, and Elliot Murphy. "Language Impairments in ASD Resulting from a Failed Domestication of the Human Brain." *Frontiers in Neuroscience* 10 (2016): 373.

Bering, Jesse M. "The Existential Theory of Mind." *Review of General Psychology* 6.1 (2002): 3–24.

Bowler, Dermot M., and Evelyne Thommen. "Attribution of Mechanical and Social Causality to Animated Displays by Children with Autism." *Autism* 4.2 (2000): 147–71.

Buber, Martin. *I and Thou*. Edinburgh: T & T Clark, 1970[1937].

Castelli, Fulvia, Chris Frith, Francesca Happé, and Uta Frith. "Autism, Asperger Syndrome and Brain Mechanisms for the Attribution of Mental States to Animated Shapes." *Brain* 125 (2002): 1839–49.

Drea, Christine M. and Alissa N. Carter. "Cooperative Problem Solving in a Social Carnivore." *Animal Behaviour* 78 (2009): 967–77.

Dunbar, Robin. *The Human Story: A New History of Mankind's Evolution*. London: Faber and Faber, 2004.
Dunbar, Robin I. M., and Julian Bever. "Neocortex Size Predicts Group Size in Carnivores and Some Insectivores." *Ethology* 104 (1998): 695–708.
Engh, Anne L., Erin S. Siebert, David A. Greenberg, and Kay Holekamp. "Patterns of Alliance Formation and Postconflict Aggression Indicate Spotted Hyaenas Recognize Third-Party Relationships." *Animal Behaviour* 69 (2005): 209–17.
Fijn, Natasha. *Living with Herds: Human-Animal Coexistence in Mongolia*. Cambridge: Cambridge University Press, 2011.
Francis, Richard C. *Domesticated: Evolution in a Man-Made World*. New York: W. W. Norton, 2015.
Freeman, Julie. "Heider and Simmel Animation." *Vimeo* video, 1:12. Posted by "Translating Nature," September 5, 2012. https://vimeo.com/48908599.
Fuentes, Agustín. "The Humanity of Animals and the Animality of Humans: A View from Biological Anthropology Inspired by J. M. Coetzee's Elizabeth Costello." *American Anthropologist* 108.1 (2006): 124–32.
Griffin, J. "Autistic Tendencies: The Consequences for Our Culture." *Human Givens* 13.4 (2006): 14–19.
Haselgrove, Mark. "Overcoming Associative Learning." *Journal of Comparative Psychology* 130.3 (2016): 226–40.
Heider, Fritz, and Marianne Simmel. "An Experimental Study of Apparent Behavior." *The American Journal of Psychology* 57.2 (1944): 243–59.
Ingold, Tim. *The Perception of the Environment: Essays on Livelihood, Dwelling, and Skill*. London: Routledge, 2000.
Klin, Ami. "Attributing Social Meaning to Ambiguous Visual Stimuli in Higher-Functioning Autism and Asperger Syndrome: The Social Attribution Task." *Journal of Child Psychology and Psychiatry* 41.7 (2000): 831–46.
Kohn, Eduardo. *How Forests Think: Toward an Anthropology Beyond the Human*. Berkeley: University of California Press, 2013.
Meyer, Matthias, Martin Kircher, Marie-Theres Gansauge, Heng Li, Fernando Racimo, Swapan Mallick, et al. "A High Coverage Genome Sequence from an Archaic Denisovan Individual." *Science* 338.6104 (2012): 222–26.
Mitchell, Robert W. *Pretending and Imagination in Animals and Children*. Cambridge: Cambridge University Press, 2002.
Murphy, Elliot, and Antonio Benítez-Burraco. "Language Deficits in Schizophrenia and Autism as Related Oscillatory Connectomopathies: An Evolutionary Account." *Neuroscience and Biobehavioral Reviews* (in press).
Nelson, Richard K. *Make Prayers to the Raven: A Koyukon View of the Northern Forest*. Chicago: University of Chicago Press, 1982.
Norenzayan, Ara, Will M. Gervais, and Kali H. Trzesniewski. "Mentalizing Deficits Constrain Belief in a Personal God." *PLoS One* 7.5 (2012): 1–8.
Nuttle, Xander, Giuliana Giannuzzi, Michael H. Duyzend, Joshua G. Schraiber, Iñigo Narvaiza, Peter H. Sudmant, et al. "Emergence of a *Homo Sapiens*-Specific Gene Family and Chromosome 16p11.2 Cnv Susceptibility." *Nature* 536 (2016): 205–29.
"Out of Shape." *YouTube* video, 4:25. Posted by University of Southern California Institute for Creative Technologies (USCICT), February 12, 2014. https://www.youtube.com/watch?v=ZAnt9II-5Co.
Reardon, Sara. "Inside the Brain of a Neanderthal." *New Scientist* 219.2926 (2013): 6–7.
Spikins, Penny. "Autism, the Integrations of 'Difference' and the Origins of Modern Human Behaviour." *Cambridge Archaeological Journal* 19.2 (2009): 179–201.
Swallow, Dallas M. "Genetics of Lactase Persistence and Lactose Intolerance." *Annual Review of Genetics* 37 (2003): 197–219.
Tishkoff, Sarah A., Floyd A. Reed, Alessia Ranciaro, Benjamin F. Voight, Courtney C. Babbitt, Jesse S. Silverman, Kweli Powell et al. "Convergent Adaptation of Human Lactase Persistence in Africa and Europe." *Nature Genetics* 39.1 (2007): 31–40.

Tomasello, Michael. "Joint Attention as Social Cognition." In *Joint Attention: Its Origins and Role in Development*, edited by Chris Moore and Philip J. Dunham, 103–30. New York: Psychology Press, 2014.

Toro, Roberto, Marina Konyukh, Richard Delorme, Claire Leblond, Pauline Chaste, and Fabien Fauchereau. "Key Role for Gene Dosage and Synaptic Homeostasis in Autism Spectrum Disorders." *Trends in Genetics* 26 (2010): 363–72.

Willerslev, Rane. *Soul Hunters: Hunting, Animism, and Personhood among the Siberian Yukaghirs*. Berkeley: University of California Press, 2007.

II

Evolving Homespun Wisdom

FOUR

Growing Wisdom

Ben Campbell

This chapter takes as its starting point the question of how the recent comparative anthropology of multi-species interactions could enliven the "evolution of wisdom" problematic. For the purpose of introducing some clarity in the terms of debates to be held, it takes issue with the human exceptionalist assumptions written into the notion of wisdom as an "inner capacity." The position to be developed in this chapter is that if wisdom is an anthropologically significant phenomenon, we can learn from the variety of ethnographic studies produced over the last twenty years about how humans interact with non-humans in distinctive places. These studies have radically overturned old Durkheimian perspectives of social morphological causality regarding nature and society, and pointed rather to the networked co-evolution of practices and ontologies in refashioning species interactions in evolutionarily consequential ways—such as domestication. Rather than "inner capacity," the wisdom of transformative patterns of interaction can be seen to involve distributed and diverse actors, human and non-human, whose aggregate roles and relations add up to more than the sum of the parts. The emergent properties of repeat practices with negative entropy loops give rise to self-organizing systems that modify and remold subjectivities, desires, and perceptions of multi-species networks. Wisdom of the extra-"inner" circumstances of being and sustenance, and of aliveness in common with other species and spirits of place, are evident in the lived worlds of foragers and horticulturalists studied by anthropologists. It is in the capacity for reciprocal attention between people and places, and the imagined connectivities of mutual co-dwelling and expectations of communicative intent, that the dynamic processes and imaginaries of social institutions for recognizing and applying wisdom have sedimented.

REVISITING DOMESTICATION

This chapter is not informed by a systematic review of the archaeology of species domestication, but puts some questions that could be useful in such an enterprise. It has benefited from a reading of James McGregor's *Back to the Garden* and his overview of human ecological relations from the archaeological record. This includes accounts of how cattle were likely domesticated from aurochs by the settlement dwellers at Çatal Huyük, where succeeding archaeological levels reveal diminishing sizes of aurochs skulls tending to tameness, and cereal grains found convenient growing conditions in the wet alluvial fan meadows of Jericho and Abu Hureyra. The book also ponders the sustainability wisdom of succeeding periods of livelihood evolution in order to reflect on the capacity of human communities to attend to the agrarian base relationships of farmlands, pastures, and settlement. My long-term interests in subsistence farming systems and environmental knowledge in the Himalayan region have taken me into "sustainability transitions" literatures, and into questions about how contemporary crises in food security projections are being addressed from a number of perspectives, including environmental justice and low carbon possibilities of food sovereignty. Among these research fields I would claim there is knowledge and creative thinking—potentially even wisdom—to be found in reflecting on multi-species interactions, for which the term "domestication" has traditionally been used. McGregor, following Andrew Sherratt, argues for origins of domestication to have been diverse, and situated in distinct ecological configurations, with the particular access of certain key sites to mountains, seas, and deserts contributing to the trans-local valuation of produce that was initially foraged, gifted, and exchanged. Only later, he argues, did this settle in, or scale up, to become a complex assemblage that we call the Neolithic.[1]

From the position of trying to think in ways that are informed by comparative anthropology, it is by way of the imaginative worlds of people whose activities create conditions of growth, and adapt to possibilities presented to them by myriad non-human species, that it might be possible to bring further perspectives to questions of evolution of wisdom. This is not a purely academic task, and the timeliness for sharing insights on the relationships outlined here, concerning prospects for sustainable growing systems, is currently much in debate in the public sphere over how we can respond to climate change.[2]

What connections do we see between people and domestication processes in the ethnographic record? Companion species imaginaries work across the intelligible registers of human-animal-plant niche interaction. There are ways that the problematic of framing wisdom in terms of navigating social networks and dynamic niches can be opened up in thinking through this nexus of species and material engagements.

SELF-ORGANIZING SYSTEMS AND ENVIRONMENTAL PERSONHOOD

Soil disturbance practices involving fire, ashes, feces, and water can be imagined on forage sites of consumption and temporary dwelling. People return and repeat, return and repeat. A pattern builds up, with the return and disturbance of the soil; the mixing of ashes again takes place, leaving foraged seeds behind.

Animals come too: pigs dig up soils for edible roots, and deer selectively browse on growing tips of preferred plants, like wild peas, beans, and barley (this happens all too frequently in my hillside garden in West Yorkshire!). Making conditions suitable for the germination of these plants that attract deer, wild sheep, and goats would be a wise move for planning to attract game, or developing the interactional dynamic with flocks of curious herbivores.

In patches of repeat dwelling, disturbance, and selective nurture of certain micro-ecologies for enhancing particular plants' emergence creates an annual pattern. This leads to congregations of people bringing foraged foods locally gathered, and given in planned and unintentional encounters, or brought from further off and contributed to a social gathering, for ritual, or for a seasonal activity such as collective fishing parties, etc. A practical nexus of consequences flows. Plants spring up in similar places, leading to the return of animals following people and vice versa. How do such networks of regularity and movement resonate in the commentaries of anthropologists, where growing patterns have found their social alignments?

Philippe Descola writes in *The Spears of Twilight* that when speaking to other French people about the horticultural system of the Achuar, he realized that in those tropical conditions there was no particularly elaborate knowledge or technical skill involved. No special explanation of cultivating knowledge was required as might be expected for growing food in, for instance, temperate European conditions. You clear a patch of forest to make a garden, put a stick of manioc in the ground, and it grows! While it might be worth warning of the tendency of slipping into the discourse of "Nature makes them lazy,"[3] there is nonetheless an important point to make here which the permaculturalists certainly follow (and which Geertz noted for the Indonesian island swidden farming systems[4]), which is the advantage of basing any food-growing system on mimicking preferential crop selection according to what kinds of plants actually want to grow in a given location. Descola writes that the manioc plants are "children" to the Achuar women who look after them. The mother plant yields children. Their men are institutionally engaged in hunting, which women cannot do. They are fundamentally who they are, as relational persons in a multi-species configuration.

The Batak in the Philippines have a notion that the rice they grow is also "human." A myth of the origin of rice tells of a father who sacrificed a child, and in order to grow rice, the Batak need to invoke the life spirit of the child in the middle of the swidden field. Rice is consubstantial with them.[5]

For the Tamang people with whom I worked in Nepal, wheat was their original crop. At the time of harvest in April–May, a small pot of the new harvest has to be given first to the clan god of the house before the people of the house may eat any of their crop. The pot is hung from the main ridgebeam of the roof. Their connection with cattle is very close. When one of their cows gives birth, all the closest members of the lineage are called to the mobile animal shelter, where they share in eating a small warm and chewy piece of the colostrum that is lightly cooked.

Tamang farming processes could be approached from an outsider's modern agronomic view and found to be inefficient in one way or another, but the actor perspective has to be held together with an etic perspective to understand why people do things and challenge the perception of what is going on, when for instance seen from children's perspectives. When asked what they were doing, they would first say that they were doing what mother or father asked them to do, before they would describe the technical task involved. This is something that anthropologists often have to learn about through fieldwork. We perhaps tend to cut out the relational context of an activity to focus on the instrumental action of human hand-eye coordination on some sticks, stones, and soils. By looking beyond the technical focus to what an agronomist-engineer would consider extraneous to the principle materials and actions in the field of interaction, we can see that what is being done is learning the process of building up the institutionally situated personal world of social being on a day-to-day basis, and this relationally lived world is what gives endurance, vitality, and resilience to the sense of being alive, flourishing, and being looked after.

In complementary function with the actor perspective, routines of activity in grooves of recurring time cycles can be assisted by happy coincidences rather than be the result of planned outcomes, which have systemic properties of multi-species interaction that are not necessarily visible to the performers. A practical nexus of consequences flows from plants springing up in places leading to the return of animals following people and vice versa.

In my fieldwork in Nepal with Tamang-speaking people, villagers would tell me that compared to maize, potatoes, and finger millet, growing wheat was "easy."[6] I pondered as to why this might be so. I thought through the sequences of technical processes of manuring, plowing, hoeing, sowing, weeding, and harvesting for the different crops. Then I realized that with wheat the weeding phase simply does not need to be done. Why is this the case? A patch of wheat yields valuable undercrops of edible weeds enjoyed by cattle over the winter. Wheat gets weeded not by "weeding" in an arable sequence of discrete tasks, but in the flow of a nexus of practices involving animals, crop plants and the voluntary "weeds of cultivation" that have evolved to grow in the spaces of a certain crop rotation, which sustains a repetitive energetic metabolism of opportunities for the needs and growing conditions of the species involved. Thinking of cattle and their well-being becomes enfolded into the needs and practices of human kinship networks, in which the daily life of domestic activity in relationally responsive obligations and exchanges involves a round of attention and care for non-human species. When daughters of the lineage marry, it is normal for them to be given a cow by their fathers and other clansmen. The givers of the cow receive copious amounts of home-distilled alcohol in return. The cows and calves themselves run skittishly around the fields and houses if they are given the fermented grain mash after distilling, happy to belong to this home.

Classificatory systems that structuralists have often found with great insight to operate between bodies, social roles, foods, and environments enable patterns of coherence to emerge and congeal around embodied practices of experience in growth processes. Christine Hugh-Jones, for example, demonstrates remarkably close correlations between ideas of bodily strength and weakness, toxically powerful and bland substances, that organize indigenous Tukanowan ways of self-understanding within an influential environment of diverse active principles and transformational materials.

In my Nepal research on environmental knowledge, it was noticeable how a certain patriarchy of authority and tribute entered into exchanges between ritual specialists—shamans (*bombo*) and founding lineage priests (*lhaben*)—and territorial deities influencing disease and behaviors of living beings. But I was frequently told by villagers about motherly spirit beings called *Kaliama*. The symbolism of the figures of the flower-dwelling *Kaliama* seemed to be different from symbolic deference to do with hierarchical and tributary gifts to territorial gods, as sovereigns of place. They seemed to concern the fecund life process itself, and the sources of generation that bring about human life and kinship.

In his textual analysis of a night-long shamanic healing, Andras Höfer comments that the chanting

> . . . repeatedly refers both to the "life-flower" and the "life-beam" of the woman patient in question. It . . . clearly indicates an organic continuity between the two, which amounts to suggesting a metonymic relation-

> ship between the "florality" and "arboreality" of the human person in general . . . The Flower as a fruit-bearing organ that grows on the branches of a tree is a key image above all in the creation myth. And the Tree/Trunk/Beam/Pole . . . is both a poetic metaphor and a ritual symbol of the backbone or the whole body in shamanic spiritual anatomy.[7]

While the *Kaliama* divine mothers are connected with flowers as symbols of procreative power and bodily health, in my understanding they are not themselves recipients of offerings, but rather belong to a personified conceptual scheme of thinking about growth, the body, and the fragility of human life in a consubstantial identity with growth processes. Speaking with one of the shaman's sons and his wife, they told me procreation is not just to do with a man's seed and a woman's womb. The intervention of *Kaliama* has to enable the physical act of conception to work. They elaborated beyond reproductive assistance to talk of the protective attendance of *Kaliama*: "One mother is not enough: she cannot look after all her children, so we need many mothers, the *Kaliama*."

There is of course, as we learn in standard treatments of Anthropology 101, a crucial way in which people are not like other species, and it is in the human community's distinctiveness from the animals' incestuous habits that separation is achieved. This puts affinity into a trajectory of personhood toward otherness and away from necessarily repulsive consubstantiality. But those communities of constituted otherness are more than frequently found to be marked by their foods, and food-related bodily effects. In the configuration of otherness, the workings of diet and formative local environments mark out communities that live by the coast/inland, up the mountain/in the plains, eating items that others hold as taboo.

Those others who do strange things with food are, however, strangely attractive in the light of needing to look elsewhere for a mate. Human life is constituted through consubstantial relational personhood spliced with the necessity of domesticating alterity through the incest taboo. This makes "us" both composite and constituted as different from those "not us" (over the ridge, in the next county who do things according to their own familial norms, and with animals that make us say "yuk, icky!").

PLACES OF CONTEMPLATION AND ENCOUNTER

The next aspect of growing wisdom to explore is the significance of place. Whether in familiar terms of *genius loci* or numinous spirits of place, comparative anthropology has come to recognize the prevalence of "sentient landscapes" that are no mere backdrop to human actions, but active fields of inter-subjectivity and encounter with otherness of human and non-human kinds. In the Tamang sensibility to landscapes, spirit forces inhabiting places are deemed responsible for chance events, fortunate outcomes in hunting, protection against harm and misfortune, landslides, destructive bolts of lightning, and keeping hailstorms from devastating one patch of crops as opposed to those of another community. It becomes hugely important to maintain communicative channels with powers in the landscape that can visit terrible consequences in responsive fashion on perpetrators of misdeeds.

Some kind of "fall"-like mythical event often marks the point of separation at which a common facility of speech between people and gods was lost. It is frequently the case that particular forest groves or lakes are places where these separations can be diminished and the worlds of gods and people brought closer, so that communication across the divide can take place. At these times, especially during pilgrimage, expressed wishes for health, long life, children, abundant crops, and livestock are routinely made.

Time and space can be enfolded in imagined causality as when the preparation of sown ground entails risks of many kinds as a new phase of growth emerges from the soil. It is frequently the case that demarcations of sacred spaces at such key turning points of the year are made in order to demonstrate respectful recognition of powers of place, who can thereby be reciprocally claimed into relationship by asking for a counter-prestation in the hope that no harm will come to the young crops. Areas of the sovereign's abode are put off-limits to all but ritually prepared entrants, and among those to be categorically excluded are transgressive types, such as forest plant foragers and menstruating women.[8] Children are from an early age taught in their rounds of play and foraging about these demarcations of dangerous places, and the vengeful acts their sovereigns will dish out.

Dealing with both randomness and human vulnerability, the first subject of conversation among the Tamang in conditions of extreme weather and destructive environmental events is the figure of "Lord of Place, Lord of Soil" (*shyibda-neda*). As the guardian of locality and living things, when landslides strike, people fall ill, or livestock disappear only to return weeks later, the interpenetration of human fates with a precocious sovereign's surveillance provides a field of problematic curiosity over cause and effect, and reflection on commission of sinful acts or human motivations. It can be not so much a punitive discourse pure and simple, but a deliberative space for putting ethical questions to the test, taking up positions from other intentions in circulation in a hunter-prey pluriverse, and hearing others' accounts of other possible causes or incidents. These are moments when wisdom and evaluative discretion can be heard. They evince a situatedness in life in the debt of giving landscapes that can just as quickly take away.

Writing of the shamanic traditions further north in inner Asia and Siberia, Morten Pedersen and Rane Willerslev discuss the points of view that are taken up in shamanic perspectives: "The animist soul is an inherently relative or deictic phenomenon, whose form depends on who perceives it and from where."[9]

The capacity noticed among hunters to position themselves through mimicry and amulets among the interactive fields of animal spirits, and their passing as beings of other sorts, plays back and forth across human, ancestor, and guardian/cannibal/possessor/trickster subjectivities in "feeding the great circle of soul exchange."[10]

Pedersen and Willerslev go on to conclude:

> Anthropologists and others with an interest in understanding non-Judeo-Christian and non-Cartesian conceptual traditions in the North Asian animist context and indeed elsewhere in the world now have ontological perspectivism as a theoretical framework allowing for conceptions of the soul that blur the distinctions between materiality and spirituality, inner and outer, self and other. Among both the Chukchi and the Darhads, as we have seen, what constitutes the invisible soul (and the visible body) depends on the perspective that one inhabits, and the two regularly trade places with each other. Indeed, the material and the spiritual become "flip sides" of one another.[11]

HORTO-REFLEXIVITY FROM FAMILIAL GROUND

As a habitual amateur gardener, I must confess to having enjoyed being useful and appreciated as a young boy for digging holes in the ground, and bashing clumps of sod into pieces with a spade. On reflection I was being "farmed out" during school holidays to spend time following my grandmother around her kitchen garden. She discovered that my imagination that the Hampshire clods of clay were enemy soldiers was no bad thing for the purpose of breaking up the heavy soil.

Over many decades I have cultivated in different places and even introduced carrots and broad beans into the Tamang village in Nepal, where I used a portion of ground on a Himalayan terrace to grow vegetables as a method for inviting observation and comment ("Ha! He plants whole potatoes, not cutting them up to make more plants. How wasteful!"). There were some "weeds of cultivation" familiar to me: dock, silverweed, amaranth (simply called "flower"—*mendo*—in Tamang), hemp (used for rope making and stomach pain medicine), and a vigorous nettle (which provides one of the key sources of food, but has to be cooked outside the house or the clan god will be offended).

With my attention having been drawn to the value of synergies between crops and animals, and the potential integration of preferential species among those plants that will willingly become weeds of cultivation, it becomes clear that if selectively weeded and allowed to seed, living communities of plant succession can appear with regularity. These contexts of interaction have probably been underreported when the narratives of domestication are told. When dominant species of domestication come to typify the process of stabilizing the cultivar for its particular traits, the enabling environments of interaction, by which emergent imaginations of accommodating selective plants in growing spaces would have come about, risk being ignored.

After eighteen years of cultivation, my current garden has abundant flushes of self-sown kale, Himalayan mustard greens, lamb's lettuce (corn salad), miner's lettuce (*claytonia perfoliata*), herbs like marjoram and sweet cicely, and the less dependably recurring, savage minded wild pansy.[12] Some of these welcome opportunists in cleared ground are now faddishly sold in high-end supermarkets as mixed leaves with rugola (in the vernacular, rocket salad), next to "ancient grain" salads. This year has seen a particularly good volunteer crop of *Chenopodium Bonus Henricus.*[13]

LANDSCAPING IMAGINED COMMUNITIES OF FOOD CROPS

In thinking about scenarios in which human-food species interactions can stabilize into regular associations of selected niche-making, the comparative ethnographic record would point toward a distributed agency of multiple species and conditions favoring frequencies of repeat movements and behaviors, and gendered human reflection on conditioning relationships for enabling sentient landscapes. Rather than the mastery of nature cosmology of contemporary crop developers, we have to imagine how landscapes would have been interacted with, and what attention might have been paid to growth, fertility, and boon for forager-cultivators. The evidence from so many cultivating communities is that treading lightly on the soils of place masters, and recognizing non-human sources of abundance, in compacts of toil and return, are more than the "naturalization" ideologies of human societies characterized by unequal ownership.

Success in food cultivation is not intrinsically different in many cultural contexts from success in hunting practices, but part of a meaningful array of places, animals, and plants, gendered domains and debts of recognition and tribute. Rituals need to mark the initiation of seasons and the bringing home of harvests. Relationships must be maintained and behavioral codes followed for good favor and possibilities of the same again next year. Group consciousness consolidates around keeping up the prospects for "we-groups" to have rights to continue taking animals and crops in legitimate ways in known territories. The giver of possibility for this is external to the group itself, and our knowledge of diverse ethnographic examples indicates that many cultures populate the landscape of sentient accountability with diverse categories of being. In the Himalayas, the shamanic terminologies make the mountain itself the deity of communicable intent and gratitude, while Buddhist versions turn the environment into a polity of rule,[14] and the Hindus evoke elemental Shiva power myths to tie in remote gushing mountain torrents with the Ganga Mata that flows into the abundant plains below. People orient their eco-imaginary worlds of locatedness to make sense of contingency and realign the forces they perceive as active to improve the chances of good fortune.

Speaking of a non-Buddhist (*bon po*) chant from eastern Bhutan, Toni Huber describes a protection ritual dealing with the lineage well-being of a royal patri-line requiring a ransom (*glud*) from a stag hunt, to exchange for the prince's life force vitality after this had been captured by demonic spirits. The "ritual antecedent narrative" deals with the search for the appropriate animal to meet the ransom demand.

> The *bon po* sits in the middle of a circle of "male dancers" (*bro pa*) and chants the *Sha ba rabs* [recitation of the stag], while the dancers perform their steps around him. This dance includes forms of animal imitation. Dancers place their hands with outstretched fingers at the sides of their heads to represent the stag's antlers, bend over at the waist, and raise their heads up to mimic the gestures of a rutting male animal, or the manner in which a stag passes through the undergrowth, so as not to snag its antlers.[15]

Huber gives this example as a deliberate contrast to the normal Buddhist deprecation of killing. The hunter moves across ridges, over and beyond one, and then over the next hill, meeting stags, but they tell him that they are stags of the wrong color. The ransom will only work with the proper colored stag, white above and red below. Then at a particular site at the margins of the meadows and the forest, the intended prey is encountered and the ransom is taken to restore the health of the prince. So it is not any stag in any place that will do. The legitimate creature in the correct location, whose whereabouts is identified by fellow stags, clinches the compact between people and provider. Huber continues that the location of "the stag's chosen place of death at the boundary between the meadow and the forest" is significant because "[i]t is precisely at such boundary places that [ransoms] and other transactions are set in motion, and dispatched to the palaces of the deities to gain favour."[16]

Clan territories are known to have been historically widespread in the central and eastern Himalayas, and among certain groups these land-people relationships were formalized through the legal principle category known as *kipat*. Collective ownership of land consisted substantially in annual ritual recognition of spiritual relationships with territory and marking the respectful bounds of ethical accommodation between humans and their needs, and other species' proper habitats and their home-worlds. This kind of landscape cannot be viewed from human-exceptionalist bias. The productive economy of subsistence forest-based agro-pastoralism is conditional upon maintaining live and active exchanges of words and things across these perspectives. Animals and plants avail themselves not through an exclusive return to labor and capital but through a cosmic economy of recognizing human and non-human others. The failure to maintain these links is what then leads to destruction through floods, hailstorms, landslides, drought (all too common in recent years of accelerated climate change), reproductive loss, and disease. The non-human world can take its revenge on human hubris-driven wealth hoarders and enact reverse flows from cultivated areas back into the forest. This is the fundamental principle that gave sustainability to early agrarian systems according to McGregor.[17]

In a discussion of the Tibetan notion of well-being Samten Karmay writes: "[T]he layman usually entreats the mountain deity to help him 'increase his fortune like the galloping of a horse and expand his prosperity like the boiling over of milk' (*rlung rta rta rgyug rgyug/ khar rje 'oma 'phyur 'phyur*)."[18]

Please note here a very similar phenomenon to the way Eduardo Kohn pays attention to semiotic resonances of actions and noises in the Amazonian jungle world.[19] The Tibetan words are very onomatopoeic in the cases of "gallop" (*rgyug rgyug*), pounding those hooves over the rangelands, and "boil" (*'phyur 'phyur*), bubbling up and expelling bursts of hot air.

CONCLUSION

In the emerging anthropology of multi-species interactions, relational markers of belonging, commitment, difference, and affinity are found to be intimately connected with companion species and notions of othering and consubstantiality among farmers and hunter-horticulturalists.

Such notions as are found among Achuar women that their garden crops are children, or among the Tamang that daughters are made into persons of adult subsistence status in the dowry of a cow given by their clansmen at marriage, have systemic consequences. These reproduce opportunities for practices depending on regular patterns of residence and dwelling. They connect environmental personhood as ongoing cultural life of daily activity with systemic effects on a community scale. At these hearths and gardens of domestic life processes, skills of culinary production and ties of cooperative relationships are fashioned in the immediate co-presences of dynamic niches. Life goes on with persons in roles that collectively make possible belonging in extended bio-social networks.

The interactional qualities of life pulse at different rates through these cycles in ways that lead people to hive off and re-gather with differing intensities. There is a strongly nomadic quality, and a characteristic feature of intentional fallowing and rest, to systems of food growing that are not driven by the productive intensification of the contemporary world. Allowing crop species to grow in favorable conditions, rather than challenging biotic system entropy, entails long cycles of soil fertility maintenance and flows from forest to fields. Enclosures of territories and social networks of distribution give rise to perceived needs for familial and clan closure that are rightly connected to the eventual emergence of property relations and regulating claims on surplus. However, relational fluidity is more important to stress for constituting the ongoing capacity of hearths to attract and give life, and to sustain communities of hearths as relationally dependent. Storage potential and redistributive networks in deferred reciprocity for exchange make extra-domestic communities of access entitlement more flexibly integrated. Adaptive institutions for long-term resilience appear from the archaeological record to be ritually elaborated, and in the ethnographic record connect disparate communities through collective dialog and negotiation with extra-human sovereigns, as in pilgrimage. Food growing communities therefore make cybernetic loops between the hearth-worlds of domestic difference, and the time and space scales of territory and seasons. These give motion and growth to hearths in their extensive and retentive practices of cooperation and boundedness, both nourished and threatened by relational being among non-human companion travellers.

The move to domestication has been argued here from comparative ethnographic evidence to involve imaginaries of sociability with crops, animals, and territories. There is no intrinsic break with hunting cosmologies that configure the world as sentient and communicative. An observance of seeds and soils, harvest and return, of season and plot pull human social imaginaries into shared worlds of interaction.

NOTES

1. James McGregor, *Back to the Garden: Nature and the Mediterranean World from Prehistory to the Present* (New Haven: Yale University Press, 2015).

2. See, for example, Pope Francis, "*Laudato Si'*: On Care for our Common Home" (May 24, 2015), http://w2.vatican.va/content/francesco/en/encyclicals/documents/papa-francesco_20150524_enciclica-laudato-si.html; Naomi Klein, *This Changes Everything: Capitalism vs. the Climate* (New York: Simon & Schuster, 2014); Jonathon Porritt, *The World We Made: Alex McKay's Story from 2050* (London: Phaidon Press, 2013); and Rob Hopkins, *The Transition Handbook: From Oil Dependency to Local Resilience* (Totnes: Green Books, 2008).

3. Mark Harris, "'Nature Makes them Lazy': Contested Perceptions of Place and Knowledge in the Lower Amazon Floodplain of Brazil," *Conservation and Society* 3.2 (2005): 461–78.

4. Clifford Geertz, *Agricultural Involution: The Processes of Ecological Change in Indonesia* (Berkeley: University of California Press, 1962).

5. Dario Novellino, "'Talking about Kaltura and Signing Contracts': The Bureaucratization of the Environment on Palawan Island (the Philippines)," in *Sustainability and Communities of Place*, ed. Carl A. Maida (New York: Berghahn, 2007), 82–108.

6. Tamang is a group of oral language dialects spoken by some 1.5 million people. See Ben Campbell, *Living Between Juniper and Palm: Nature, Culture, and Power in the Himalayas* (Delhi: Oxford University Press, 2013).

7. Andras Höfer, *A Recitation of the Tamang Shaman in Nepal* (Bonn: VGH Wissenschaftsverlag, 1994), 311.

8. Campbell, *Living Between Juniper and Palm.*

9. Morten Pedersen and Rane Willerslev, "'The Soul of the Soul Is the Body': Rethinking the Concept of Soul through North Asian Ethnography," *Common Knowledge* 18.3 (2012): 467.

10. Ibid., 469.

11. Ibid., 486.

12. Matthew Hall writes: "evidence in the plant sciences is increasingly recognizing the capacity of plants to assess, perceive, and act on their environment. The primary purpose of their doing so is to generate the conditions for their own flourishing. With these plant purposes in mind, it is evidentially inappropriate to treat plants purely as instruments. In this age of environmental crisis, knowledge of plant intelligence and autonomy opens up a new debate on respecting and promoting the well-being of the plants that make life on Earth possible." Matthew Hall, "Plant Autonomy and Human-Plant Ethics," *Environmental Ethics* 31.2 (2009): 169.

13. That great source of wisdom *The Modern Herbal*, vol. 1 by Mrs. M. Grieve (New York: Dover, 1971[1931]) lists its common names as Good King Henry, English mercury, mercury goosefoot, allgood, tola bona, smearwort, and fat hen. She writes "The leaves used to be boiled in broth, but were principally gathered, when young and tender, and cooked as a pot herb. In Lincolnshire, they are still eaten in place of spinach. Thirty years ago [i.e., 1900], this Goosefoot was regularly grown . . . in Suffolk, Lincolnshire, and other eastern counties, and was preferred to the Garden Spinach, its flavour being somewhat similar, but less pronounced. In common with several other closely allied plants, it was sometimes called 'Blite' (from the Greek, *bliton*, insipid). Evelyn says in his *Acetaria*, 'it is well-named being insipid enough.' Nevertheless, it is a very wholesome vegetable. If grown on rich soil, the young shoots, when as thick as a lead pencil, may be cut . . . and boiled and eaten as Asparagus. They are gently laxative." Grieve, *The Modern Herbal*, 365.

14. Brigitte Steinmann, "Mountain Deities, the Invisible Body of the Society: A Comparative Study of the Representations of Mountains by the Tamang and the Thami of Nepal, the Lepcha and Bhotia of Sikkim," in *Reflections of the Mountain: Essays on the History and Social Meaning of the Mountain Cult in Tibet and the Himalaya*, ed. Anne-Marie Blondeau and Ernst Steinkellner (Wien: Verlag der Österreichischen Akademie der Wissenschaften, 1996), 179–218.

15. Toni Huber, "Hunting for the Cure: A Bon Healing Narrative from Eastern Bhutan," in *Tibetan and Himalayan Healing*, ed. Charles Ramble and Ulrike Roesler (Kathmandu: Vajra Books, 2015), 372.

16. Ibid., 375.

17. McGregor, *Back to the Garden*.

18. Samten Karmay, "The Wind Horse and the Well-Being of Man," in *Tibetan and Himalayan Healing*, ed. Charles Ramble and Ulirike Roesler (Kathmandu: Vajra Books, 2015), 408.

19. Eduardo Kohn, *How Forests Think: Towards an Anthropology Beyond the Human* (Berkeley: University of California Press, 2013).

BIBLIOGRAPHY

Campbell, Ben. *Living Between Juniper and Palm: Nature, Culture, and Power in the Himalayas*. Delhi: Oxford University Press, 2013.

Descola, Philippe. Spears of Twilight: *Life and Death in the Amazon Jungle*. London: Harper Collins, 1966.

Geertz, Clifford. *Agricultural Involution: The Processes of Ecological Change in Indonesia.* Berkeley: University of California Press, 1962.

Grieve, Margaret. *The Modern Herbal.* New York: Dover, 1971[1931].

Hall, Matthew. "Plant Autonomy and Human-Plant Ethics." *Environmental Ethics* 31.2 (2009): 169–81.

Harris, Mark. "'Nature Makes Them Lazy': Contested Perceptions of Place and Knowledge in the Lower Amazon Floodplain of Brazil." *Conservation and Society* 3.2 (2005): 461–78.

Höfer, Andras. *A Recitation of the Tamang Shaman in Nepal.* Bonn: VGH Wissenschafts-verlag, 1994.

Hopkins, Rob. *The Transition Handbook: From Oil Dependency to Local Resilience.* Totnes: Green Books, 2008.

Huber, Toni. "Hunting for the Cure: A Bon Healing Narrative from Eastern Bhutan." In *Tibetan and Himalayan Healing,* edited by Charles Ramble and Ulrike Roesler, 371–412. Kathmandu: Vajra Books, 2015.

Hugh-Jones, Christine. From *The Milk River: Spatial and Temporal Processes in Northwest Amazonia.* Cambridge: Cambridge University Press, 1979.

Karmay, Samten. "The Wind Horse and the Well-Being of Man." In *Tibetan and Himalayan Healing,* edited by Charles Ramble and Ulrike Roesler, 403–10. Kathmandu: Vajra Books, 2015.

Klein, Naomi. *This Changes Everything: Capitalism vs. the Climate.* New York: Simon & Schuster, 2014.

Kohn, Eduardo. *How Forests Think: Toward an Anthropology Beyond the Human.* Berkeley: University of California Press, 2013.

McGregor, James. *Back to the Garden: Nature and the Mediterranean World from Prehistory to the Present.* New Haven: Yale University Press, 2015.

Novellino, Dario. "'Talking about Kaltura and Signing Contracts': The Bureaucratization of the Environment on Palawan Island (the Philippines)." In *Sustainability and Communities of Place* , edited by Carl A. Maida, 82–108. New York: Berghahn, 2007.

Pedersen, Morten, and Rane Willerslev. "'The Soul of the Soul Is the Body': Rethinking the Concept of Soul through North Asian Ethnography." *Common Knowledge* 18.3 (2012): 464–86.

Pope Francis, "*Laudato Si'*: On Care for our Common Home." May 24, 2015. http://w2.vatican.va/content/francesco/en/encyclicals/documents/papa-francesco_20150524_enciclica-laudato-si.html.

Porritt, Jonathon. *The World We Made: Alex McKay's Story from 2050.* London: Phaidon Press, 2013.

Steinmann, Brigitte. "Mountain Deities, the Invisible Body of the Society: A Comparative Study of the Representations of Mountains by the Tamang and the Thami of Nepal, the Lepcha and Bhotia of Sikkim." In *Reflections of the Mountain: Essays on the History and Social Meaning of the Mountain Cult in Tibet and the Himalaya,* edited by Anne-Marie Blondeau and Ernst Steinkellner, 179–218. Wien: Verlag der Österreichischen Akademie der Wissenschaften, 1996.

FIVE

Homo Sapiens Sapiens

The Human as Homemaker

Julia Feder

Jesuit priest and theologian Avery Dulles (1918–2008) has described contemporary Catholic theological anthropology as turning on the insight that "[t]o be a human . . . is to be a body-person, an incarnate spirit."[1] This way of framing human existence, with an emphasis on embodied spirituality, is meant to push against dualistic anthropologies (both in the Christian tradition and in contemporary popular imagination) that portray an irresolvable tension between the flesh and the spirit. He continues, "To come into one's own as a person is, under one aspect, to become related, through the body, to a surrounding world. Religious awareness, paradoxically, requires a turning *to* the world; for only in a spiritual movement toward finite realities can one actuate the sense of the transcendent as that which goes *beyond* the world."[2] If (Christian) "religious awareness" or "com[ing] into one's own as a person" consists in becoming consciously aware of one's embodied modes of relationality, then we can claim in a related fashion that Christian wisdom must involve embodied modes of paying attention to the interrelated nature of created reality. This includes, inherently, an embodied attention to one's material context. This mode of description of Christian wisdom is not meant to imply an acceptance of the world as it is as the way it must always be, but rather implies an attention to the dynamic nature of reality and one's own role (and responsibilities) in the unfolding of history. Thus, Christian wisdom, in "turning *to* the world" (as Dulles puts it), involves an invitation to conscious participation in created reality as co-creators with God: we are invited to carve out a niche; we are invited to convert space into place and place into home.

We find a similar (if not unexpected) emphasis on the importance of material context in recent developments of evolutionary theory today, especially in the emergence of the extended evolutionary synthesis.[3] Proponents of the extended evolutionary synthesis highlight niche construction as an integral component of the evolutionary process: organisms are not only affected by their environmental contexts, but they also actively create and destroy ecological niches.[4] Relative to other organisms, human beings are particularly active niche constructors via a suite of symbolic-social behavior. Notably, the human converts space into place in and through her symbolic-social capacities. This process of conversion is a constructive process constituted equally by constraints and facilitations, whereby living into material constraints opens up new imaginative possibilities and, in turn, new imaginative possibilities open up new material realities. If it is true, as some contemporary evolutionary anthropologists suggest, that a fundamental component of human behavioral modernity is the conversion of space into place,[5] and it is also true, as religious historian Thomas Tweed argues, that the work of religion is "homemaking,"[6] then the Christian tradition as a school of wisdom should offer its adherents some models of wise homemaking, or "good housekeeping," if you will.

In this chapter, I will take human niche construction as a starting point for reflection on Christian wisdom as homemaking. In the first half of this essay I argue that Christian theology needs to take account of the ways in which time and place are not mere impositions upon human being (as perhaps twentieth-century Catholic theologian Edward Schillebeeckx's inclusion of conditioning by time and space among his "anthropological constants" might suggest), but instead that time, space, and human beings interact in a dynamic process of "mutual mutability."[7] In the second half of the chapter, I offer a portrait of Jesus as homemaker—viz., as one who imaginatively creates and facilitates a home within the material constraints of his social and historical location. As one who both embodies the fullness of humanity and incarnates divine Wisdom, this reading of Jesus is able to facilitate a robust Christian reading of human wisdom as homemaking—i.e., simultaneously living in tune with the way the world is and actively constructing it as the kind of place in which we desire to live.

Contemporary Christian theology, as Avery Dulles suggests above, has widely incorporated non-dualistic anthropologies as a baseline for contemporary orthodoxy. This has been facilitated by the turn to history in contemporary social theory starting in the nineteenth century, and has been given particular weight in Christian theology in the mid-twentieth century (see, e.g., the work of Johannes Baptist Metz, Wolfhart Pannenberg, Jürgen Moltmann, and Edward Schillebeeckx). An acknowledgment of the importance of history and the embodied nature of both epistemology and salvation has reigned. Yet, the significance of place is an aspect of a fully embodied theology that has been undertheorized. Though Michel Foucault predicted that this period of intellectual history in which we are currently living will be later remembered as the "epoch of space," Christian theologians have yet to participate in cultural theory's "recent 'spatial turn.'"[8] Although many theologians have given attention to the ways in which diverse human cultural contexts and experiences condition theological truth[9] and some have thought through the effect of discourses of place on Trinitarian theology,[10] epistemology,[11] and human experience,[12] Christian theologians have yet to reflect thoroughly on the theological significance of the human being as the kind of animal that actively makes places and, in so doing, encounters God.[13]

The writings of Edward Schillebeeckx, especially insofar as he embodied a "turn to history" in the later part of his career, represent the state of contemporary Christian theology well in this respect. That is to say, Schillebeeckx's writings reflect both the ways that Christian theologians have taken material context seriously and the ways that theologians largely have yet to integrate in a full way attention to place and place-making into Christian thought. For Schillebeeckx, epistemology is fully conditioned by history, experience, and the limits of the body. His maxim, "No salvation outside the world," represents an embodied and historical notion of Christian salvation—Christian salvation is at once material, interpersonal, social-political, and spiritual. Yet Schillebeeckx does not adequately address place and place-making as a significant component of human interaction with the material world. However, there are aspects of Schillebeeckx's thought that can offer some points of departure for a theological spatial turn: namely, his articulation of the conditioning of "space" as shaping human flourishing and, even more significantly, his portrait of Jesus as one who operated creatively within the limits of his time and space. With a fresh attention to the significance of place-making in human behavior and meaning-making, we can utilize these aspects of Edward Schillebeeckx's thought as starting points for developing a fuller portrait of the ways that time, space, and human beings interact in a dynamic process of "mutual mutability,"[14] and in particular how Christian wisdom involves (good) homemaking.

THE EVOLUTION OF THE HUMAN AS HOMEMAKER

The extended evolutionary synthesis, as a narrative of evolutionary processes which includes both non-genetic and non-selection processes as causal evolutionary forces, can give us a framework for considering place as a dynamic co-agent. A key element of the extended evolutionary synthesis is niche construction theory, a theory that emphasizes the "capacity of organisms to modify their environment and thereby influence their own and other species' evolution."[15] Therefore, organisms are more than simply reactive to their environments—they also actively construct them through the building and destroying of niches.[16] Organisms and their environments are thus "mutually mutable" and "synergistic."[17] Many organisms are niche constructors (including even earthworms, whose landscape altering activity Darwin studied), yet the extent of human niche construction is unparalleled.[18] Thus, biologists and anthropologists frequently refer to the human as the "ultimate niche constructor."[19] As Jeremy Kendal explains, human niche construction proceeds largely via socially transmitted (rather than genetically transmitted) behaviors. These behaviors may have an ultimate effect on the human genome or they may not. Regardless of their genetic impact, they remain a significant component of human evolutionary inheritance.[20] Notably, the social construction of human spaces as structured learning environments creates the conditions for the possibility of apprentice learning,[21] identified by anthropologists as significant for human behavioral modernity.

Behaviorally modern human beings have carved out a complex cognitive, behavioral, and ecological landscape in which all is shot through with semiotic meaning. Human patterns of interaction with each other, with other creatures, and with landscapes are thoroughly conditioned by symbolic interpretations of reality.[22] Therefore, space itself, for humans, is never experienced as empty but rather it is always infused with socially-constructed forms of meaning.[23] Space thus becomes "place" as it is inhabited by humans.[24] In this process of habitation, a place becomes densely, complexly, fragilely, and fluidly[25] packed with semiotic meaning, viz. human wisdom. In other words, a place is given order and history. It becomes the bearer of a future. Just as humans transform space into place, places mutually transform humans into persons. Places are the spaces that hold interpersonal (and inter-species) relationships together and form communal identity. Places give humans a history and the possibility of a future. Humans and place are mutually conditioning.[26] Place-making is such a basic human activity that cultural and linguistic anthropologist Keith Basso writes, "As roundly ubiquitous as it is seemingly unremarkable, place-making is a universal tool of the historical imagination. And in some societies at least, if not in the great majority, it is surely among the most basic tools of all."[27]

Theologians, historians of religion, and social theorists construct placial theories to reflect the significance of place-making in human cognition and behavior. For these theorists, "place" is symbolic space, constructed by the human though not untethered to material reality. As Christian theologian Sarah Morice-Brubaker explains, "place happens . . . when a thinking subject seeks to organize and endow with meaning a world that she finds herself already in and part of. Environment and subject, together, enact place."[28] Thus, place-making is not an exercise in pure human caprice—it is a creative activity of "endow[ing] the found world with meaning, directionality, and navigability."[29] This imaginative play is undergone in relation to the real constraints and possibilities presented by the world as it is experienced. Yet, at the same time, space, for humans, is always "much more than just material, geometric extension . . . Under the auspices of place, one's belonging, one's identity in relation to one's surroundings, is at issue . . . there to be negotiated, and in fact already in the process of being negotiated."[30] Thus, places are symbolic spaces in which reality is co-determined by the material and the non-material.[31]

Space as place is inhabited by *persons*. That is, place is an inherently relational space and place-making is (communal/social) identity-making. As Basso describes,

> . . . what people make of their places is closely connected to what they make of themselves as members of society and inhabitants of the earth, and while the two activities may be separable in principle, they are deeply joined in practice. If place-making is a way of constructing the past, a venerable means of *doing* human history, it is also a way of constructing social traditions and, in the process, personal and social identities. We *are*, in a sense, the place-worlds we imagine.[32]

Place-making and its attendant human activity—identity-making—is never the work of isolated human individuals but is instead the work both of human communities and of humans in relation to other organisms in their niche.[33] As Ben Campbell suggests in his essay in this volume, place is always a space of encounter or "mutual co-dwelling,"[34] or as Marcus Baynes-Rock puts it, "human communities are never only human."[35] If the conventional axiom, as most of us have memorized it, is that no two objects can occupy the same place at the same time,[36] then perhaps we tend to have it backward: no place, in the technical sense of the term "place," ever holds only one object—place is the space of relation between objects and organisms, simultaneously present, past, and future.[37] Thus one can argue in a related fashion, as Tweed does, that space, as place, is simultaneously *differentiated* and *kinetic*.[38] Space as a place is differentiated insofar as human space is always experienced and mediated through the specific, particular and localized.[39] Space as place is kinetic insofar as space changes and moves with the passage of time. Theologian Mary McClintock Fulkerson dramatizes this reality as she writes about the kinetic nature of the "hometown," at once fluid and yet also unified:

> Provocations of many-layered memories serve up new versions of the hometown at different periods. What "gathers" to make place is unavoidably fluid. Thus, the boundaries of place are always permeable, modulating over the years and continuing to change. When we visit our hometown, we could even say that we find it a *new* place. On the

> other hand, we always know there is a clear *something* referred to by this place we call hometown, even if only because we say it is no longer "the same place."[40]

To be in a place is to be in the temporal (and relational) flow of history.

Places are both *generated* and *generative*, meaning that places "come into being through embodied [human] perception, figurative imagination, and ritual practice"[41] and they, in turn, come to exert an agency of their own.[42] As Tweed suggests, "We might say playfully: [places] are people too."[43] Landscapes and inhabitants form a symbiotic relationship in which both give shape and meaning to each other in a dynamic of mutual feedback. Human communities render spaces meaningful, converting them into places and from places into home. In turn, places render their human inhabitants meaningful, giving shape to identity and providing both the material and the semiotic resources necessary for community life.[44] Place-making is as much about the makers as the made. Or, to put it differently, organisms and landscapes form each other mutually. The makers and the made are continually shifting and reversing. The term "home" then dramatizes this relationship. Because "home" is where a person[45] lives, it calls attention to the ways in which one cannot think of place without place-maker and one cannot think of landscape without inhabitant.

Tweed argues that religions have functioned for human beings, at least partly, as traditions of place-making. He explains, "[Religion] is homemaking. In other words, as clusters of dwelling practices, religions orient individuals and groups in time and space, transform the natural environment, and allow devotees to inhabit the worlds they construct."[46] In other words, religions are traditions of finding one's (collective) place in space and actively participating in place-making so as to "intensify joy and confront suffering."[47] If Tweed is right that religions orient devotees in space and construct meaning via practices of homemaking,[48] then how does the Christian tradition involve dwelling practices? In other words, how does the Christian tradition function as a store of wisdom about good practices of coming home to one's body, preserving and sanctifying the domestic space, inhabiting one's region, and finding one's place in the cosmos? Does the Christian tradition offer its devotees any practices of wise homemaking?

Although the "placial turn" within contemporary critical theory has yet to make a significant impact on Christian theology, contemporary Christian thinkers have begun to take seriously the role of the body and the role of history in theological thought. Because the body is the most intimate sphere of homemaking and because an awareness of historical context is integral to any project of beginning to "locate" universal claims, these are important components to acknowledging the central position of place-making in human thinking about the divine. Yet more needs to be done. Edward Schillebeeckx's writings on the parameters of salvation reflect well both the best of contemporary Christian thinking on history and materiality and the limitations of contemporary Christian reflection on place-making, so I will turn to him as a case study so as to lay bare the state of the question in mainstream contemporary Christian thought.

EDWARD SCHILLEBEECKX AND PLACE

In his 1977 text (published in English translation in 1989), *Christ: The Experience of Jesus as Lord,* Edward Schillebeeckx outlines seven "anthropological constants" in order to specify the parameters of human flourishing (which he identifies positively with Christian salvation). It is significant that Schillebeeckx explicitly includes the conditioning effect of "space" in his fourth anthropological constant, fully titled, "the conditioning of people and culture by time and space."[49] For Schillebeeckx, space figures prominently in determining the *limits* of human flourishing: if, for him, there is "no salvation outside the world,"[50] there is also no salvation outside of the particularities of the world which one inhabits—including, specifically, the "historical" and "geographical" particularities of one's context.[51] In his discussion of this anthropological constant, he offers little on the positive contributions of space and place to human salvation. His point in including a discussion of time and space in his list of anthropological constants is to push against a tendency prevalent in the Christian and Western intellectual tradition to offer abstract, universal claims, and instead to argue with force that we are always and everywhere culturally and historically situated. Humans are meaning makers in the context of their own space and place. Earlier in the same text he writes,

> The historicity and thus the finality of man, which he does not know how to escape so that he can adopt a standpoint outside time, makes him understand his humanity also as a *hermeneutical* undertaking, i.e., as a task of *understanding* his own situation and unmasking critically the meaninglessness that man brings about in history.[52]

Therefore, we interpret (and indeed construct) meaning perpetually because of the reality that we are situated. There can be no once-and-for-all construction of meaning that is not subject to revision and reformulation.

If Schillebeeckx's discussion remains largely negative in the fourth anthropological constant, perhaps we can look to one of his earlier anthropological constants as a clue to some more positive ways in which place shapes human well-being. As Schillebeeckx frames his first anthropological constant, "The relationship of the human being to his own corporeality—man *is* a body but also *has* one—and by means of his own corporeality to the wider sphere of nature and his own ecological environment, is constitutive of our humanity. So human salvation is also concerned with this."[53] Thus, Schillebeeckx names our bodies and our ecologies as overlapping spheres of physical locatedness significant for human flourishing. These are our layered "household[s]."[54] Yet even here he continues to use negative language, e.g., "Our relationship with nature and our own corporeality come up against *boundaries* which we have to respect if we are to live a truly human life and, in an extreme instance, if we are even to survive. Therefore, what is technically possible has not by a long way been an ethical possibility for men which makes sense for them and to which they can respond."[55] Despite the fact that Schillebeeckx does not give explicit attention to the positive role of place and place-making in human life, Schillebeeckx's writings can begin to set a framework for the consideration of homemaking as an integral component of human flourishing (and, consequently, of human salvation). Though he does not use the term "homemaking" himself or explicitly reflect on placemaking as a human practice, his attention to the multiplicity of ways in which humans are located and locate themselves (e.g., historically, geographically, in a body, in an ecology) resonates well with contemporary placial theorists and can serve as an adequate starting point for thinking through a complex understanding of Christian wisdom as homemaking. Even more promising, however, is his understanding of the life, death, and resurrection of Jesus. Schillebeeckx describes Jesus as one who carved out a place for himself and those he loved in the context of his own particular constraints and facilitations. Schillebeeckx's Jesus is a homemaker, an incarnation of Divine Wisdom herself.

JESUS AS HOMEMAKER

In the Jewish scriptural tradition that the early Christians inherited, Wisdom is a personification of God's creative and saving action in the world.[56] Wisdom is operative from the moment of creation, bringing order and beauty to the cosmos, just as "place" is ordered and live-in space (Proverbs 3, 8). She is portrayed as one who stands in the streets in the middle of the day and preaches in a loud voice, inviting all into relationship with her (Prov. 1:20–33; 8). In preparation, she builds a columned house, cooks meat, mixes wine, and spreads a feast upon a table. She sends out messengers to relay her message of invitation. She is a ready and generous host, providing a place and a home for all who love her (Prov. 9). She warns that acceptance of this invitation means life and refusal means death (Prov. 1, 4; Sirach 4). Alienation from her means alienation from the land itself, literally displacement (Prov. 2:21–22). In another set of images similarly evoking language of home and placemaking, Wisdom surveys all of creation searching for a place to dwell, and she chooses to make her home in Jerusalem. Here she pitches a tent and strikes her "root[s]" in order to grow as a tall tree, providing perfume and rich fruits from her flowering branches (Sirach 24). Wisdom commands justice toward one's neighbors and those under one's care (Prov. 3, Sirach 4). In and through her own homemaking activities, she invites all to imitate her hospitality and generosity. Lovers of wisdom are to be homemakers, generously bringing God's order, justice, and beauty to their own domains.

As the early Christians tried to make sense of the theological significance of Jesus, they turned to Wisdom imagery as a resource. Borrowing directly from the Wisdom traditions of the Hebrew Scriptures, Paul portrays Christ as operative in the creative activity of God and as the one who holds all of creation together (Colossians 1:15–23; 1 Corinthians 8:6),[57] even, at times, explicitly identifying Christ as "wisdom" (1 Cor. 1:30). The author of the gospel of Matthew redacts his source Q in order to transfer quotations from Sophia to the mouth of Jesus.[58] And the author of the gospel of John begins his text with a classic wisdom hymn. Throughout all four gospels Jesus is described as one who preaches justice and hospitality and who initiates a community of fellowship organized around the common table.

The dominant interpretation of Jesus for Schillebeeckx is that of the "eschatological prophet"[59] who initiates eschatological salvation by embodying it in his manner of life. For Schillebeeckx, Jesus (like Sophia) was a "prophet of salvation,"[60] announcing that God's cause is the human cause[61] and, thus, that God both desires and promises comprehensive human good. Specifically, Jesus' manner of life was characterized by his "caring and abiding presence among people experienced as salvation coming from God."[62] His very presence was experienced as facilitating individual and social flourishing. In particular, as a table companion, his presence brought others freedom and deep joy.[63] As Schillebeeckx writes, "Being sad in Jesus' presence [was] an existential impossibility."[64] Commenting on the distinction in the gospel of Mark between the asceticism of John the Baptist and the comparatively lavish eating and drinking of Jesus and his companions, Schillebeeckx explains, "if Jesus' disciples should fast now, that would be to disregard the concrete situation, namely, the presence of salvation in the actual person of Jesus of Nazareth. The disciples' being together with this Jesus is in essence a festive celebration of good fellowship, a meal prepared by Jesus himself, and a fellowship in which there is salvation."[65]

In the Christian tradition, salvation is imaged in two related ways and both images are brought into focus in Jesus's practice of sharing food. Therefore, I highlight Jesus' practice of food-sharing as an expression of Sophia's own homemaking practices. Sophia's homemaking practices included food preparation and food sharing as primary gestures of hospitality and the embodiment of God's saving love for all. The first image, the resurrection of the body, represents the personal-material well-being that Christians look forward to with eschatological salvation, and the second image, the Kingdom of God, represents the social-political well-being that Christians anticipate. These images work together to indicate the holistically mystical-political shape of Christian salvation.[66] Though the second image, the Kingdom of God, is more obviously highlighted in the practice of food-sharing, particularly to the degree that the gathering together to a common table those who have been socially alienated represents the restoration of social and political right relationship, the hope for personal-material well-being symbolized in the resurrection of the body also has a role to play. The practice of sharing food nourishes individual material bodies just as it restores social and political bodies. Even the resurrected body desires food and delights in eating with others (Luke 24:41–43). In other words, food sharing serves as a rich symbol of a broader understanding of salvation (both personal and social-political) in the Christian tradition. For this reason, I use food sharing to stand symbolically for the broader whole of human homemaking activities that make up both individual and social forms of placial meaning-making.

The joy and freedom experienced at table with Jesus functions as an ecclesial hermeneutic: to be church, to participate in the Christian community, is to eat and drink with Jesus. In particular, Jesus' table fellowship with those who were normally considered beyond God's hospitality and nurturing care—Gentiles, sex workers, those working in collaboration with Roman imperial authorities to oppress the Jews, etc.—symbolizes God's own openness to friendship with all of humanity.[67] As Schillebeeckx explains, "in Jesus' earthly career and ministry we are seeing demonstrated the praxis of the Kingdom of God which he preached and implemented. In his earthly, historical life the eschatological praxis of the coming rule of God is already coming into view within the dimension of our human history here on earth . . . in this concrete Jesus-phenomenon, proclamation, praxis and person cannot, it seems, be separated."[68] The fellowship meal, then, is not simply an icon of the Kingdom of God. It is a *real symbol* of the Kingdom of God, i.e., a sign that brings about that which it signifies. As Schillebeeckx writes, "this fellowship at table is itself, as an eating together with Jesus, an offer here and now of eschatological salvation or 'final good.'"[69] And, consequently, "the call of Jesus seems to have a fundamental connection with being a table-companion,"[70] an openness to this transformative food-sharing that is, itself, an instantiation of the Kingdom of God.

Jesus' practice of establishing community around a common table participates in the reality of eschatological salvation in a real, yet self-consciously partial way as it anticipates the fullness of completion in the promised future. Therefore, the wholeness that Jesus announced is still to come, but, at the same time, it is also already present (in germ) in the gathering of friends around a single table.[71] As Schillebeeckx explains, "Jesus proclaims the salvation to come, and at the same time by his conduct he makes it present."[72]

When viewed in light of the scriptural tradition of Sophia as homemaker (a move that Schillebeeckx does not explicitly make himself), Jesus can be seen as one who embodies Sophia's homemaking activities with such an integrity that his message and his person are one: he announces the Kingdom of God as a placial orientation (though not yet fully realized) and also converts his space into this place (in an "already" fashion) through his homemaking activity. Home, then, is both "already" present in the process of homemaking itself, yet also always "not yet" for Jesus and his followers.

Despite his status as a wandering healer and prophet, Jesus repeatedly enacts the role of generous host throughout the gospel narratives, thus embodying the divine activity of Sophia. During his life, Jesus' feeding of others is significant: all of the gospels include a narration of Jesus' miraculous feeding of many, and some (Mark and Matthew) even repeat the story with another version.[73] Schillebeeckx argues that the healing and nature miracles of Jesus were recounted in the gospels to dramatize (regardless of particular historical accuracy) Jesus' way of life as an embodiment of God's concern for humanity and God's desire for right relationship.[74] Specifically, Schillebeeckx argues that the miraculous feeding stories function in the gospels to "condense a basic feature of Jesus' manner of offering fellowship during his days on earth."[75]

In the face of a death sentence at the hands of the political authorities, Jesus was forced to make sense of his circumstances, especially in light of his prior experience of God as Abba (i.e., God as one who cherishes humanity and desires human liberation). As Schillebeeckx puts it, Jesus "was forced by circumstances to give a place to approaching death in his radical confidence in God."[76] In the face of this forced placial meaning-making, Jesus returns to a familiar activity—he gathers his friends together for a meal and plays the role of both host and servant. Because Schillebeeckx recognizes the destructive consequences of soteriologies that valorize suffering as a means to redemption, he is careful to note that one cannot claim that

> Jesus both willed and sought after his death as the sole possible way of realizing the kingdom of God. There would have been an element of play-acting about his commitment to his message of *metanoia* and the rule of God, if he had thought and known from the very start that salvation would come only in consequence of his death.[77]

Yet, given that his death was inescapable and resulted from the integrity of living out his mission, "Jesus had to settle for himself what his attitude to impending death must be."[78] As the gospels narrate it, Jesus enacted this meaning-making and spoke directly about his impending death at a farewell meal, thus situating the interruptive and utterly negative event of his death within the more intentional and positive arc of his life and mission.[79] At this meal he signaled to his followers that the fellowship initiated in this space cannot be destroyed with his death. Whenever his friends come together and remember him, he will be present with them. In other words, "the 'Jesus affair' is to go ahead" in the face of execution.[80] This was Jesus' way of avoiding "passively allowing death to overcome him"[81] and transforming an unavoidable event into an opportunity to extend his homemaking activity. Thus, Jesus' death is now able to function at the service of bringing about the Kingdom. This is not because the Kingdom of God requires the spilling of blood, but rather it is due to Jesus' own wise negotiation within the constraints and facilitations of his context. He does this negotiation with the faith that God is capable of redemption in even the most impossible of circumstances. Though Schillebeeckx does not use this term, we can argue that this kind of wise homemaking—this carving out a place—is the very condition for the possibility of Christian salvation.

With a fresh attention to the significance of place-making for humans, we can understand Jesus' activities in the face of death (and the early Christian interpretation of the resurrection) as a kind of placial construction. As Mary Magdalene and the other women[82] come upon the tomb of Jesus and find it emptied, they understand (gradually) that this is not the place where one can encounter Jesus after his death.[83] In contrast to the positive "place" of the last supper—where Jesus gives his body to his friends and declares that this is the authentic "place" of (continued) meaningful encounter—the empty tomb functions as "no place."[84] Death has lost its sting (1 Cor. 15:55, Hosea 13:14); the tomb has been emptied. Instead, it is the shared table that is the meaningful space of encounter. The "real place," so to speak, is not the grave but the table. Post-resurrection, both at Emmaus and on the shore, his disciples recognize him only after he has acted as host. In Luke, he blesses bread and breaks it in another's house, and they recognize him as the one who died and was risen (24:30–35).[85] In John, on the shore of the Sea of Tiberias, Jesus prepares a breakfast feast of fish and bread to share with the disciples (21:9–14). He converts whatever space is available into a place of his own to host others. Though he "has no place to lay his head" (Matthew 8:20; Luke 9:58) he makes a home and invites others to share around a table wherever he finds himself. Thus, Jesus' homemaking is not a kind of placial construction untethered from material reality or somehow less dependent on physicality than other forms of place-making. It is representative of the very kind of place-making that humans are apt to perform—i.e., processes of meaning-making in which the subject constructs meaning within and together with the constraints and facilitations of one's (physical, social, political) environment. For Jesus, the kind of place that he constructs is explicitly a communal or socially-constructed place and this contributes to its resiliency. This is paradigmatic of the wisdom of human place-making. More specifically, with the resurrection of Jesus, which interestingly is not narrated as an event itself in the Gospels,[86] a "collective—ecclesial—experience" is given birth.[87] The resurrection makes possible a communal place centered on the collective experience of the risen One. This place is made manifest in the assembling together of his friends and their homemaking activity. Violence cannot dislocate this place that Jesus has constructed. The way that Jesus has lived his life has planted him so firmly within these relationships of love that no one can displace him.

CONCLUSION

Place-making is a creative human task. Place neither impinges upon us wholly from without (i.e., it does not function as a mere limit), nor is it wholly something of our own making. Place-making proceeds in the face of both constraints and facilitations. In the Christian tradition, if we are attentive to homemaking as a fundamental human activity, Jesus can provide a model of wise homemaking, especially insofar as he actively participates in the construction of meaning of his life and death. The table is the "place" of meaning—the hermeneusis of both his life and his death and the place where his friends encounter him together. His tomb remains an empty space—no place for him. In the shared meal all spheres of "home" come into focus (body, domestic space, homeland, and cosmos) and his followers and friends find orientation. The family meal, an item of domestic life, instantiates (in germ) both the Kingdom of God, i.e., the homeland, as well as the resurrection of the body which Christians hope to experience more fully with the eschatological transformation of the cosmos. Jesus' followers experience, in this practice, (however partially) a coming home to their bodies and to their world, as Dulles suggests. Jesus' persistent and creative homemaking practices illustrate the dynamic breadth of human placial wisdom at its best.

Yet, an abstract discussion of Jesus as homemaker can only do so much work. A theoretical account of the significance of place-making for humans and for the divine-human relationship has its limits, though it points us in the right direction by focusing us on the community located in a specific place. The theological thought of Edward Schillebeeckx can offer us a framework for insisting that place is an anthropological constant, despite the reality that he does not discuss place as much more than a negative conditioning element. His thought can also offer us a starting point for an interpretation of Jesus' life, death, and resurrection as homemaking activities, though he does not use the lens of homemaking himself. Yet, if we want to take seriously place as generative and differentiated, then we can only understand Jesus' homemaking activities as a basic starting point. This can provide us with a theoretical framework for considering the importance of place and human place-making, but not yet a consideration of how Christian wisdom is placial with any specificity. If wisdom consists in turning to the world (as Dulles puts it), it is not just any abstracted world, but *this* world—the particular constraints and facilitations of my place and yours. Schillebeeckx's insight that the homemaking activity of Jesus continues today in the community of those gathered in his name does more to focus our attention and flag a task than it does to provide concrete pathways for Christian homemaking. This practical work needs to be the work of differentiated communities committed to inhabiting space and open to allowing place to generate new material possibilities.

If Jesus was one who created a place for himself and his friends, his project of homemaking is yet unfinished. We are still in the process of coming home to our bodies and our ecologies. There is a need for local Christian communities to prioritize homemaking as a mystical-political task and to arrive at specific traditions and practices sensitive to their own place(s). It is time not only to abandon the Enlightenment idea that we come from nowhere, but also to proclaim positively that wisdom can be practiced in committing to a place (i.e., not only being from a place and allowing oneself to be shaped by it, but also committing to transforming a place).[88] As Basso has argued, wisdom does sit in places. Christian wisdom is placial. It is co-constructed by communities and landscapes.

NOTES

1. Avery Dulles, "The Symbolic Structure of Revelation," *Theological Studies* 41.1 (1980): 59.
2. Ibid., 59–60.
3. Kevin N. Laland et al., "The Extended Evolutionary Synthesis: Its Structure, Assumptions and Predictions," *Proceedings of the Royal Society B: Biological Sciences* 282 (2015): 1–14.
4. F. John Odling-Smee, Kevin N. Laland, and Marcus W. Feldman, *Niche Construction: The Neglected Process in Evolution* (Princeton: Princeton University Press, 2003).
5. See, for example, Agustín Fuentes, "What Evolution, the Human Niche, and Imagination Can Tell Us about the Emergence of Religion," *Theology Today* 72.2 (2015): 171.
6. Thomas A. Tweed, *Crossing and Dwelling: A Theory of Religion* (Cambridge, MA: Harvard University Press, 2006).
7. Agustín Fuentes, "Human Evolution, Niche Complexity, and the Emergence of a Distinctively Human Imagination," *Time and Mind* 7.3 (2014): 244.
8. Tweed, *Crossing and Dwelling*, 9.
9. See, for example, James Hal Cone, *A Black Theology of Liberation* (Philadelphia: Lippencott, 1970) and Gustavo Gutiérrez, *A Theology of Liberation: History, Politics, and Salvation* (Maryknoll: Orbis, 1973).
10. See Sarah Morice-Brubaker, *The Place of the Spirit: Toward a Trinitarian Theology of Location* (Cambridge: James Clarke, 2013) for a spatialized discussion of God.

11. See Mark Wynn, *Faith and Place: An Essay in Embodied Religious Epistemology* (Oxford: Oxford University Press, 2009) for a discussion of the effect of religiously differentiated places on faith practices and religious epistemology.

12. See David Brown, *God and the Enchantment of Place: Reclaiming Human Experience* (Oxford: Oxford University Press, 2004) for an argument that the Christian historical tradition of valuing "place" as sacramental can be a launching point for contemporary reenchantment.

13. One should note there are several Christian theologians who have begun to reflect on the human as one who makes places, yet clearly, more work needs to be done. In particular, ecological theologians frequently reflect on the earth writ large as our home; see, for example, Ernst Conradie's *An Ecological Christian Anthropology: At Home on Earth?* (New York: Ashgate, 2005). In addition, Philip Sheldrake writes about the human as one who makes cities and argues that urban landscapes should be designed to reflect more than narrowly defined market concerns—they should also be designed as portals to the sacred. See Philip Sheldrake, *The Spiritual City: Theology, Spirituality and the Urban* (Chichester: Wiley-Blackwell, 2014). Similarly, Timothy J. Gorringe reflects on the everyday built environment as a locus for theological ethics. Timothy J. Gorringe, *A Theology of the Built Environment: Justice, Empowerment, Redemption* (Cambridge: Cambridge University Press, 2002). In this essay, I offer a reflection that in comparison to these works is both more abstract (especially insofar as I do not limit my reflection to a particular kind of human space—i.e., urban or rural) and also more particular (insofar as I reflect on the life, death, and resurrection of Jesus as a model for Christian dwelling practices).

14. Fuentes, "Human Evolution," 244.

15. Jeremy Kendal et al., "Human Niche Construction in Interdisciplinary Focus," *Philosophical Transactions of the Royal Society* 366 (2011): 785.

16. Ibid., 786.

17. Fuentes, "Human Evolution," 244.

18. There are virtually no species or landscapes that have not been touched by human niche construction. See, for example, Benjamin S. Halpern et al., "A Global Map of Human Impact on Marine Ecosystems," *Science* 319 (2008): 948–52 or Oscar Venter et al. "Sixteen Years of Change in the Global Terrestrial Human Footprint and Implications for Biodiversity Conservation," *Nature Communications* 7.12558 (2016): 1–11, who argue that 75 percent of the planet's land surface is currently experiencing measurable threats to biodiversity due to human behavior and that this percentage is "rapidly increasing" (Venter, "Sixteen Years," 2). Positively, it is possible for us to construct niches with attention to the future of our species as well as other species with compassion and prudence.

19. Jeremy Kendal, "Cultural Niche Construction and Human Learning Environments: Investigating Socio-Cultural Perspectives," *Biological Theory* 6.3 (2011): 241–50; Kevin N. Laland, John Odling-Smee, and Sean Myles, "How Culture Shaped the Human Genome: Bringing Genetics and the Human Sciences Together," *Nature Reviews: Genetics* 11 (2010): 137–49; Fuentes, "Human Evolution."

20. Kendal, "Cultural Niche Construction," 241–50.

21. Kendal, "Human Niche Construction," 787. Kim Sterelny, *The Evolved Apprentice: How Evolution Made Humans Unique* (Cambridge, MA: MIT Press, 2012).

22. As Agustín Fuentes puts it, "During our evolutionary history humans developed a cognitive and behavior process wherein experiences in, and perceptions of, the world exist in a context in which the 'material' world is never without semiotic meaning . . . For human beings, even early ones, their social relationships, landscapes, and the biotic and abiotic elements they encounter are embedded in an experiential reality that is infused with a consistent potential for meaning derived from more than the material substance and milieu at hand. This results in a distinctive way of being in, and perceiving of, the world for humans relative to other mammals and primates, and even other hominins" ("Human Evolution," 243). See also Terrence Deacon, *The Sym-*

bolic Species: The Co-Evolution of Language and the Brain (New York: W. W. Norton, 1997).

23. Religious historian Thomas Tweed argues that, for humans, "'space' is not a preexisting static container isolated from other spaces. Space is not an absence waiting for a presence. It's not undifferentiated vacuity, a void to be filled. It's not static or outside the flow of time. It's not self-contained or beyond the flow of causes." Instead, Tweed argues that space is always and everywhere "differentiated, kinetic, interrelated, generated, and generative." Thomas Tweed, "Space," *Material Religion* 7.1 (2011): 117.

24. Because of our propensity for organizing our world(s) with meaning, space qua space is a second order level of abstraction for humans. This kind of abstraction, of course, is enabled by the very same symbolic capacities that construct space into place in the first place. Thus, humanistic geographer Yi-Fu Tuan can describe "space" and "place" as an irreducible conceptual dyad where "space is an extrapolation from place, and place requires an incipient assumption about space" (Morice-Brubaker, *Place of the Spirit*, 29). We can never think of pure "space" without some dependence upon our experience of place.

25. Mary McClintock Fulkerson, *Places of Redemption: Theology for a Worldly Church* (Oxford: Oxford University Press, 2007), 36. In other words, semiotic meaning is both multi-layered and kinetic. See below for a more full discussion of both of these aspects of semiotic placial meaning.

26. Fuentes, "Human Evolution," 244.

27. Keith H. Basso, *Wisdom Sits in Places: Landscape and Language Among the Western Apache* (Albuquerque: University of New Mexico Press, 1996), 5.

28. Morice-Brubaker, *Place of the Spirit*, 6.

29. Ibid., 34.

30. Ibid., 1. This is a fundamental tenant of "postmodern geography." See Edward Casey, *The Fate of Place: A Philosophical History* (Berkeley: University of California Press, 2013), 197–201.

31. Basso, *Wisdom Sits in Places*, 75.

32. Ibid., 7.

33. Keith Basso writes, "it is simply not the case, as some phenomenologists and growing numbers of nature writers would have us believe, that relationships to places are lived exclusively or predominately in contemplative moments of social isolation. On the contrary, relationships to places are lived most often in the company of other people" (ibid., 109), and, I would add, in the company of other organisms as well.

34. Ben Campbell, "Growing Wisdom," this volume.

35. Marcus Baynes-Rock, "Lions, Cows and Gods," this volume.

36. This is a popular extension of the Pauli exclusion principle of quantum mechanics used in everyday communication to explain a range of phenomena from car crashes to the pitfalls of multitasking.

37. Edward Casey, "How to Get from Space to Place in a Fairly Short Stretch of Time," *Senses of Place*, ed. Steven Feld and Keith Basso (Santa Fe: School of American Research Press, 1996), 13–52.

38. Tweed, "Space," 120.

39. Ibid., 118. Tweed notes helpfully that particularized, localized places need not be limited to those that are sensually encountered, since "the never-experienced contours of a desired terminus (Christ's heaven or Amida's Pure Land) or a distant territory (Jerusalem or Mecca) might orient daily life and elicit affective response more than tangible sites nearby" (ibid., 119). These imagined places, though not sensually experienced in the usual way we think of sensory experience, are just as real, specific, and generative as those experienced with the senses. Perhaps even distinguishing between sensually experienced places and non-sensually experienced places obscures the deeply embodied relationship we have with places, even in their absence. Imagined places often do generate sensory experiences; they both condition our senses and enter our experience via the senses. For example, the sixteenth-century Spanish

Christian mystic Teresa of Avila writes about the experience of longing for the beatific vision (expected in the place "heaven," you might say) as a sensory experience in her account of pain after encounter with God in prayer. See, for example, Teresa of Avila, "The Book of Her Life," in *The Collected Works of St. Teresa of Avila*, trans. Kieran Kavanaugh and Otilio Rodriguez, vol. 1 (Washington, D.C.: Institute of Carmelite Studies, 1976), 177 and 251–52. Here sensory experience is an experience of the pain of dislocation or displacement rather than the experience of positively inhabiting a place, all the more dramatized by Teresa's own language of her bones and joints being pulled *apart*.

40. McClintock Fulkerson, *Places of Redemption*, 30.

41. Tweed, "Space," 121.

42. Though I rely upon Keith Basso significantly in this piece, I disagree with some of his assumptions about the relationship of humans and place, particularly insofar as he can seem to sometimes fail to leave room for the co-agency of landscapes in human-landscape place-making. For example, he writes, "Human constructions par excellence, places consist in what gets made of them—in anything and everything they are taken to be—and their disembodied voices, immanent though inaudible, are merely those of people speaking silently to themselves" (Basso, *Wisdom Sits in Places*, 109). Instead, I would favor a description of the "mutual mutability," to use Agustín Fuentes's language, between humans and landscapes. Landscapes impose authentic constraints and possibilities that shape the imaginative work that takes place therein.

43. Tweed, "Space," 122.

44. Basso, *Wisdom Sits in Places*, xiv.

45. I use the term "person" intentionally here to indicate a relational entity.

46. Tweed, *Crossing and Dwelling*, 82.

47. Ibid., 54.

48. Here, "home" is meant to refer broadly to at least four layers of placial dwelling: the body, the domestic space (in contrast to the public space of a community), regional space (what Tweed calls the "homeland"), and the cosmos.

49. Edward Schillebeeckx, *Christ: The Experience of Jesus as Lord*, trans. John Bowden (New York: Crossroad, 1980), 738.

50. Edward Schillebeeckx, *Church: The Human Story of God*, trans. John Bowden (New York: Crossroad, 1990), 12.

51. Schillebeeckx, *Christ*, 738.

52. Ibid., 739.

53. Ibid., 734.

54. Ibid.

55. Ibid.

56. Biblical scholars are not in agreement about the status of Wisdom in the Hebrew Bible. Some argue that Wisdom is the personification of one attribute of God or of the wisdom available to devotees after spiritual training. Some of this disagreement stems from a genuine diversity of portraits of Wisdom across various canonical and deutero-canonical texts. Other aspects of this disagreement may stem from the biases of scriptural exegetes who are uncomfortable with associating feminine imagery too closely with YHWH, as Elizabeth Johnson argues in "Jesus, the Wisdom of God: A Biblical Basis for Non-Androcentric Christology," *Ephemerides Theologicae Lovanienses* 61 (1985): 275. Here, I favor Johnson's own interpretation of the wisdom tradition: a functional equivalence of Sophia and YHWH.

57. See also Hebrews 1:3 for a similar view, though probably not written by Paul.

58. See Johnson, "Jesus, the Wisdom of God," 280-81 for a concise description of this.

59. Edward Schillebeeckx, *Jesus: An Experiment in Christology*, trans. Hubert Hoskins (New York: Crossroad, 1979), 174, 177, 274–82.

60. Ibid., 139.

61. Edward Schillebeeckx, *On Christian Faith: The Spiritual, Ethical and Political Dimensions* (New York: Crossroad, 1987), 21–22.

62. Schillebeeckx, *Jesus*, 179.
63. Ibid., 200.
64. Ibid., 201.
65. Ibid., 203.
66. Edward Schillebeeckx, "Erfahrung und Glaube," in *Christerlicher Glaube in moderner Gesellschaft*, ed. F. Böckle, F.-X. Kaufmann, K. Rahner, and B. Welte, vol. 25 (Freiburg: Herder, 1980), 99f; Schillebeeckx, *Jesus*, 141.
67. Schillebeeckx, *Jesus*, 206.
68. Ibid., 213.
69. Ibid., 218.
70. Ibid., 204.
71. Ibid., 50.
72. Ibid., 152.
73. Ibid., 213.
74. Ibid., 179–200.
75. Ibid., 214.
76. Ibid., 302.
77. Ibid., 306.
78. Ibid., 306.
79. Ibid., 307–8.
80. Ibid., 311.
81. Ibid.
82. While the three synoptic gospels have other women as Mary Magdalene's companions, the Gospel of John narrates Peter and "the other disciple" as her companions at the tomb.
83. Schillebeeckx, *Jesus*, 336–37.
84. Schillebeeckx affirms that the empty tomb functions as a negative symbol (ibid., 336). The story is not meant to document the historical event of finding an emptied tomb, but rather a symbol of resurrection faith—he clarifies, "even if this be a historical fact, theologically it could yield no proof of a resurrection. A 'vanished corpse' is not in itself a resurrection, and an actual bodily resurrection does not require as its outcome a vanished corpse" (ibid., 381).
85. Ibid., 216.
86. Ibid., 331.
87. Ibid., 352.
88. This might have implications for rethinking the nature of Christian witness. This work has already begun. See, for example, the vast body of literature on "new evangelization" in Catholic circles or, in a non-Catholic context, Anita Hooley Yoder, "A Mission to Themselves: Changing Views of Mission in North American Mennonite Women's Organizations," *Anabaptist Witness* 3.2 (2016): 17–28.

BIBLIOGRAPHY

Basso, Keith H. *Wisdom Sits in Places: Landscape and Language Among the Western Apache*. Albuquerque: University of New Mexico Press, 1996.

Brown, David. *God and the Enchantment of Place: Reclaiming Human Experience*. Oxford: Oxford University Press, 2004.

Casey, Edward. *The Fate of Place: A Philosophical History*. Berkeley: University of California Press, 2013.

———. "How to Get from Space to Place in a Fairly Short Stretch of Time." In *Senses of Place*, edited by Steven Feld and Keith Basso, 13–52. Santa Fe: School of American Research Press, 1996.

Cone, James H. *A Black Theology of Liberation*. Philadelphia: Lippencott, 1970.

Conradie, Ernst. *An Ecological Christian Anthropology: At Home on Earth?* New York: Ashgate, 2005.

Deacon, Terrence W. *The Symbolic Species: The Co-Evolution of Language and the Brain.* New York: W. W. Norton, 1997.
Dulles, Avery. "The Symbolic Structure of Revelation." *Theological Studies* 41.1 (1980): 51–73.
Fuentes, Agustín. "Human Evolution, Niche Complexity, and the Emergence of a Distinctively Human Imagination." *Time and Mind* 7.3 (2014): 241–57.
———. "What Evolution, the Human Niche, and Imagination Can Tell Us about the Emergence of Religion." *Theology Today* 72.2 (2015): 170–81.
Gorringe, Timothy J. *A Theology of the Built Environment: Justice, Empowerment, Redemption.* Cambridge: Cambridge University Press, 2002.
Gutiérrez, Gustavo. *A Theology of Liberation: History, Politics, and Salvation.* Maryknoll: Orbis, 1973.
Halpern, Benjamin S. et al. "A Global Map of Human Impact on Marine Ecosystems." *Science* 319 (2008): 948–52.
Hooley Yoder, Anita. "A Mission to Themselves: Changing Views of Mission in North American Mennonite Women's Organizations." *Anabaptist Witness* 3.2 (2016): 17–28.
Johnson, Elizabeth. "Jesus, the Wisdom of God: A Biblical Basis for Non-Androcentric Christology." *Ephemerides Theologicae Lovanienses* 61 (1985): 261–94.
Kendal, Jeremy. "Cultural Niche Construction and Human Learning Environments: Investigating Socio-Cultural Perspectives." *Biological Theory* 6.3 (2011): 241–50.
Kendal, Jeremy et al., "Human Niche Construction in Interdisciplinary Focus." *Philosophical Transactions of the Royal Society* 366 (2011): 785–92.
Laland, Kevin N., John Odling-Smee, and Sean Myles. "How Culture Shaped the Human Genome: Bringing Genetics and the Human Sciences Together." *Nature Reviews: Genetics* 11 (2010): 137–49.
Laland, Kevin N., Tobias Uller, Marcus W. Feldman et al. "The Extended Evolutionary Synthesis: Its Structure, Assumptions and Predictions." *Proceedings of the Royal Society B: Biological Sciences* 282 (2015): 1–14.
McClintock Fulkerson, Mary. *Places of Redemption: Theology for a Worldly Church.* Oxford: Oxford University Press, 2007.
Morice-Brubaker, Sarah. *The Place of the Spirit: Toward a Trinitarian Theology of Location.* Cambridge: James Clarke, 2013.
Odling-Smee, F. John, Kevin N. Laland, and Marcus W. Feldman. *Niche Construction: The Neglected Process in Evolution.* Princeton: Princeton University Press, 2003.
Schillebeeckx, Edward. *Christ: The Experience of Jesus as Lord.* Translated by John Bowden. New York: Crossroad. 1980.
———. *Church: The Human Story of God.* Translated by John Bowden. New York: Crossroad, 1990.
———. "Erfahrung und Glaube." In *Christerlicher Glaube in moderner Gesellschaft,* edited by Franz Böckle, F.-X. Kaufmann, K. Rahner, and B. Welte, vol. 25, 73–116. Freiburg: Herder, 1980.
———. *Jesus: An Experiment in Christology.* Translated by Hubert Hoskins. New York: Crossroad, 1979.
———. *On Christian Faith: The Spiritual, Ethical and Political Dimensions.* New York: Crossroad, 1987.
Sheldrake, Philip. *The Spiritual City: Theology, Spirituality and the Urban.* Chichester: Wiley-Blackwell, 2014.
Sterelny, Kim. *The Evolved Apprentice: How Evolution Made Humans Unique.* Cambridge, MA: MIT Press, 2012.
Teresa of Avila. "The Book of Her Life." In *The Collected Works of St. Teresa of Avila.* Translated by Kieran Kavanaugh and Otilio Rodriguez, vol. 1. Washington, D.C.: Institute of Carmelite Studies, 1976.
Tweed, Thomas A. *Crossing and Dwelling: A Theory of Religion.* Cambridge, MA: Harvard University Press, 2006.
———. "Space," *Material Religion* 7.1 (2011): 116–23.

Venter, Oscar et al. "Sixteen Years of Change in the Global Terrestrial Human Footprint and Implications for Biodiversity Conservation." *Nature Communications* 7.12558 (2016): 1–11.

Wynn, Mark. *Faith and Place: An Essay in Embodied Religious Epistemology*. Oxford: Oxford University Press, 2009.

III

The Wisdom of Speech

SIX

Speaking Truthfully

A Thomistic Perspective on the Peculiar Origins of Human Language

Stewart Clem

Human beings are liars. While most people would hesitate to classify themselves as such, the fact remains that we tell lies all the time, often without a moment's hesitation. Even those of us who avoid telling malicious falsehoods will admit to any number of "harmless" or "white" lies: we bend the truth in order to avoid hurting someone's feelings, to avoid responsibilities, and to make excuses. Of course, there is some variation across cultures in terms of just how acceptable or even expected lying might be in given contexts,[1] but no one disputes that the deliberate telling of falsehoods is a near universal feature of human existence. Lying is a learned behavior, and children typically begin to display a capacity for lying in social situations around 3–4 years of age.[2] Yet, if a person wished to commit themselves to the complete avoidance of telling lies, it would require an almost unthinkable degree of self-determination. As theologian Paul Griffiths observes, absolute non-liars "almost always bear the burden of unremitting honesty together with some clinical psychological disorder: they're retarded, insane, or saintly—or perhaps all three at once."[3]

Moreover, lying is a peculiarly human activity. While non-human animals are capable of deception, humans are the only species capable of telling lies. There is no universally recognized definition of lying, of course, either across cultures or among theorists. While we might quibble about the definition of lying, the fact is that we have never discovered anything like "lying" (on most definitions, at least) in non-human animal behavior.[4] In this chapter, I consider lying as a distinctly human phenomenon. Recognizing that its definition is disputed, for our present purposes I will define lying as the utterance (or writing) of a statement that one believes to be false. This definition has held great purchase in the history of Western philosophy, and it was shared by Augustine of Hippo and Thomas Aquinas.[5] Writing in the fourth century, Augustine once observed, "Indeed, nothing should be thought more peculiar to humankind than lies and falsehoods, which derive from humanity itself."[6] More recently, some anthropologists have even posited that lying is the primary function of human language.[7]

If lying is a marvel of human language, we may rightly ask whether it is a good thing or a bad thing. Lying—in order to be executed successfully, at least—requires a certain degree of cleverness. But is it a reflection of human *wisdom*? Moreover, is it simply a linguistic expansion of the practice of deception (which we *do* find in animal behavior)? It should be noted that my proposed definition of lying does not require intent to deceive. But surely there is some connection between lying and deception. In light of these considerations, my aim in this chapter is to examine the role of deception in the evolution of language as it relates to normative claims about lying and truth-telling in human societies. My research draws upon contemporary analyses, but I am primarily interested in how these analyses intersect with the thought of the scholastic theologian Thomas Aquinas, whose position on lying represents one of the most influential in the history of Western thought. In his analysis of lying, Aquinas states that "spoken words are by nature signs of one's thoughts,"[8] and uses this claim to support his argument that lying is never permissible.[9] On its surface, the argument seems be that since language has a particular natural function (i.e., communicating the contents of one's mind), and lying is a violation of that function, then lying is always wrong. Many readers of Aquinas are not quite sure what to make of this. In what follows, I suggest that an examination of the role of deception in the evolution of human language—from perspectives in anthropology, comparative psychology, and philosophy—gives us a better grasp of what it might mean to say, along with Aquinas, that lying is a violation of the natural function of language. In the process, it will become clear that lying is in fact a manifestation of human wisdom, insofar as we are able to categorize and make normative claims about it.

NORMATIVITY AND THE "NATURALNESS" OF LANGUAGE

Aquinas is often criticized for grounding his argument against lying (at least in part, it seems) in the idea that uttering falsehoods violates the natural function of speech. Even those who wish to defend the thesis that lying is never permissible frequently dismiss this aspect of his thought. Christopher Tollefsen, in his recent monograph, *Lying and Christian Ethics*, laments,

> This looks very much like a version of a perverted faculty argument, an argument that holds that the "natural function" of something should not be thwarted deliberately. Such arguments are hardly compelling in their most common setting, which concerns the use of bodily organs, such as sex organs. That the natural function of an organ *is* such and such does not provide a *reason* for agents to respect that function, at least, not absent some account of the relation of the function to a good that is preserved or promoted by that function. Thus, on their own, natural-function arguments are empty of their normative significance.[10]

This objection sums up nicely the general resistance one often finds to Aquinas' prohibition of lying. Tollefsen is articulating a version of the "Is-Ought" fallacy, made famous by eighteenth-century British philosopher David Hume.[11] Should we be concerned, along with some of Aquinas' interpreters, that he has mistakenly derived an "ought" from an "is"? To make matters even worse, Tollefsen explains,

> The argument that Aquinas seems to be giving in this passage is made more problematic by its seeming focus on the natural purpose of *language*, for language is an artifact, something made by human beings to serve their purposes; this is something St. Thomas is quite aware of. And there is no principled reason why an artifact made to serve one purpose might not have several purposes.[12]

Thus, not only has Aquinas incorrectly inferred a morally normative claim directly from a feature of the natural world, he seems to have momentarily forgotten that language is not even a "natural" phenomenon at all—it is an *artifact*, a human creation. Tollefsen still finds Aquinas' analysis of lying to be persuasive on the whole, but he rejects this specific line of argument as an embarrassing flaw. In order to rescue Aquinas, Tollefsen draws our attention away from language and into the broader domain of *communication*. He believes that this is Aquinas' fundamental concern in classifying truth as a virtue, and from this observation he infers that the act of lying is always opposed to one of two basic goods: *the good of society* and *the good of personal integrity*.[13] These interpretive moves allow him to read Aquinas as making moral claims that are not grounded in any sort of natural function.

Tollefsen is right, of course, to draw attention to the virtue of truth. This is the context in which Aquinas' analysis of lying is situated. Yet, it seems to me that Tollefsen has not adequately considered the ways in which the natural functions of language permeate Aquinas' entire analyses of truth and lying. More importantly, I do not think that he has adequately considered language as a natural phenomenon, nor the significance of truth-telling as a constitutive feature of language. My primary purpose here is not to single out Tollefsen's analysis for critique, but I find that his arguments represent those of a number of Aquinas' interpreters on the problems of language and the ethics of lying. Another relevant concern, and also frequently overlooked, is the fact that Aquinas situates his own arguments against lying within the broader context of the virtue of truth. Aquinas writes, "Now there is a special order according to which our outward actions, whether words or deeds, are duly ordered to things, just as "sign" is ordered to "thing signified." In this way a person is perfected by the virtue of truth."[14] Any analysis of Aquinas' position on lying, therefore, must give due consideration to the virtue of truth. But what does it mean for truth to be a virtue?

It is a common assumption that virtue ethics proposes an alternative to an ethics of obligation. While this may be true in one sense, it would be misguided to assume that the virtues do not generate moral norms or duties. This is made manifestly clear when we consider the virtue of justice. For Aquinas, the virtue of justice is "a habit according to which a person, with a constant and perpetual will, renders to each that which is his or her right."[15] Justice refers both to the equitable treatment of persons and to the specific virtue cultivated in an individual.[16] Given this definition, we should not be surprised to find that Aquinas classifies truth as a virtue annexed to justice. Truth brings with it an obligation, a moral debt, so to speak. This means that, in certain contexts, we owe other persons a manifestation of the truth.[17] We may not always be obligated to speak, but when we do speak we are obligated to speak truthfully. This normative claim is grounded in certain aspects of human nature: "Words are by nature signs of one's thoughts" and therefore "it is unnatural and undue that anyone should signify with words that which is not in one's mind."[18] As I have already noted, many people remain unpersuaded by such arguments, and it is not always clear whether Aquinas thinks that the natural characteristics of human speech in and of themselves generate definite normative claims. We may find Aquinas' remarks here incomplete or unsatisfying, but this is likely a position he and his contemporaries would have taken for granted. Moreover, as is frequently the case with Aquinas, there are implied intermediate premises in his argument that render it more subtle than might appear at first glance. We must also keep in mind that there are underlying pragmatic concerns driving the argument. For example, he indicates that truth is necessary for the preservation of human society; as he explains, "It would be impossible for humans to live together, unless they believed one another, as declaring the truth to one another."[19] It is worth asking whether this is meant to be a new, distinct argument, or whether it somehow overlaps with the "naturalness" argument.

For the moment, I want to leave these concerns and questions looming in the background. Turning now to some contemporary theorists, I want to shift our focus to the problem of deception in human evolution. As scholars in various disciplines have noted, human language depends on mutual trust in order to function at all.[20] Some have hypothesized that the practice of deceptive vocal utterances has in fact hindered other species from developing the skills required for language.[21] Yet, humans overcame this barrier and developed sophisticated speech practices such as narrative and displaced reference.[22] What does this tell us about human beings? Moreover, can it help us to articulate a better formulation of Aquinas' natural normativity of language?

The writers I engage in the following sections represent a variety of academic disciplines. In no way do I mean to suggest that their work reflects a consensus across these disciplines, or even that they are motivated by the same set of concerns. Nor do I wish to suggest that any individual scholar is representative of his or her own field. Qualifications aside, however, there *are* certain claims that have found widespread acceptance among scholars in both the biological and the social sciences, and these often serve as starting points for more advanced hypotheses in their own projects. One such emerging consensus is the view that the phylogenetic development of human language cannot be explained simply by the emergence of more complex "brain hardware" in hominins, nor solely through a process of favorable genetic mutations.[23] Perhaps the most influential figure who has contributed to this (now maligned) view of language is linguist, philosopher, and cognitive scientist Noam Chomsky. Throughout his career, Chomsky developed the notion of a "universal grammar," or the basic conceptual structure for all human languages, that is embedded in the physical structures of the human brain. This model, which Chomsky coined the "language acquisition device," was meant to explain why very young children are able to learn their native languages so rapidly.[24] As will become clear in what follows, this view has been, if not refuted, then at the very least, complicated by newer theories in multiple disciplines.[25]

INSIGHTS FROM ANTHROPOLOGY

Anthropologist Chris Knight was among the early challengers to the theory that human speech evolved as a result of more advanced biological hardware. Drawing on the work of Robin Dunbar[26] and others, Knight proposed a framework for the development of speech that made sociality and joint intentionality—in other words, *behavioral* and not merely *biological* factors—integral to the evolutionary development of language itself, through a process he coined as "ritual/speech coevolution."[27] On Knight's account, human language is a revolutionary development brought about by social cooperation. Central to Knight's framework is the notion of "conspiratorial whispering." Other animal species developed means of communication through ingroup formation, such that the cost of sharing information was low, yet highly valuable. This sharing of information is valuable both for the species and for the individual, and the value is obvious enough to encourage the kind of cooperation necessary to spread useful information to members of the group. Human speech meets the criteria of conspiratorial whispering, but Knight observes that the development of language is still perplexing given that our primate ancestors lacked the mental capacity for language. Moreover, the possibility of deception precludes the development of language in primate species because it is too costly, in evolutionary terms. *Language*, as opposed to other forms of communication, opens up new possibilities for deception. How did humans overcome this barrier and develop what we would now recognize as language?

In order for speech to develop, Knight argues, not only must the barrier of deception be overcome, but *honesty* must become the default of human speech. He explains:

> We humans rely not merely on unintended truthfulness in one another's signals: where we are on speaking terms, we expect intentional honesty. It follows that without the establishment among humans of a new kind of honesty as a default—habitual honesty in volitional signaling—speech could not have got off the ground. In the human case, then, precisely the most unreliable kinds of signals—namely, the volitional, intentional ones—must have become adapted for honest use. Somehow, in the course of human evolution, what were once frequency-dependent tactical deceptions must have become increasingly routine while becoming simultaneously harnessed to a reversed social function—the group-wide sharing of good information.[28]

The only way for this to happen, Knight insists, was for "sufficiently intense ingroup trust" to develop, igniting the possibility of speech on all fronts. That is, ingroup trust coalesced with biological and behavioral capacities that were already developed. According to Knight,

> The problem, then, is that conventional signals depend on trust, whereas those animals intelligent enough to use such signals will also be clever enough to exploit that trust competitively. This may help explain why, despite their cognitive capacities . . . chimpanzees have no natural use for conventional signals. In particular, it clarifies why, in common with other primates, chimps do not vocalize dispassionately, lacking those capacities for cortical control which appear natural in other contexts such as manual gesticulation . . . Such lack of control should not be seen as maladaptive: at stake is the maintenance of credibility. Chimps, like other primates, need reliable signals on which to base their behaviour. Only to the extent that their vocalizations remain governed by the limbic (emotional) system can listeners trust them as reliable cues to internal states.[29]

Knight thinks that if sufficient ingroup trust can be established, then the problem of lying goes away, since liars become ostracized by the group. He gives an account of how this could have happened, which, while still speculative, draws on his decades of research on the role of ritual in human evolution, with a special emphasis on courtship rituals. In contrast to scholars like Chomsky and Derek Bickerton, Knight argues that language did not evolve as a result of more sophisticated brain hardware; rather, it began with sharing in representational activity—specifically, representational activity that arose in the context of rituals. His proposal suggests that, rather than developing in piecemeal fashion, the various components of language—metaphor, displaced reference, duality of patterning and syntax[30]—are logically interrelated and came into being as a unified whole.[31] These developed simultaneously through selection pressures against deceptive signals (especially sexual signals), leading to solidification of trustworthy vocal utterances. This model, in addition to addressing the problem of deception, also helps to explain a number of related problems in linguistics, from tense/case markers to syntactical competence (i.e., the ability to recognize and process "syntactical complexity," such as modality and conditional statements).[32] As Knight acknowledges, we still need a complete evolutionary story to make sense of these claims, but his hypotheses are theoretically sound and accord with current evolutionary biology. Another upshot of Knight's framework is that it enriches the idea that human beings are, in the classical sense, "social animals" in that their most distinctive capacities—rationality and linguistic communication—are bound up in communal life.

This model helps us to see more clearly that language did not arise in a vacuum, and that human sociality is as much a cause of language as it is a result of it. Terrence Deacon, an anthropologist who has written at length on the co-evolutionary processes that facilitated humans' use of symbolic reference and language, writes, "It is simply not possible to understand human anatomy, human neurobiology, or human psychology without recognizing that they have all been shaped by something that could best be described as an idea: the idea of symbolic reference."[33] Like Knight, Deacon proposes that the use of symbolic reference first arose in the context of ritual practices. Unlike Knight, however, his research is less concerned with the problem of deception than with the broader problem of *why* symbolic communication arose in the first place. Deacon avers that, from the perspective of hindsight, symbolic communication is "overdetermined," in the sense that everything in our evolutionary history seems to point to it.[34] Thus, while we need not be concerned about discovering a monolithic explanation for its origins, by the same token we cannot afford to ignore the complex role that sociality has played in human evolution. "Though symbolic thinking can be entirely personal and private," explains Deacon, "symbolic reference itself is intrinsically social . . . Our uniquely human minds are, in a very concrete sense, the products of an unusual reproductive challenge that only symbolic reference was able to address—a concrete internalization of an ancient and persisting social evolutionary predicament that is uniquely human."[35] Thus, while it is true that language is the greatest facilitator of human sociality, it is equally true that language could first arise only in a species comprising social animals.[36]

INSIGHTS FROM COMPARATIVE PSYCHOLOGY

Chomsky's model of the language acquisition device is not the only commonly held assumption in linguistics to come under suspicion. As Jean-Louis Dessalles has argued, the "classic" explanation for the existence of human language, namely, that it benefits the species as a whole, is simply wrong.[37] While there may have been a time when the evolutionary advantages of language seemed obvious, this is no longer the case. Scientists, linguists, and anthropologists have noted too many difficulties and complicating factors in language to assume that the development of language is an inevitable destination in the evolutionary trajectory of a certain lineage of primates. More precisely, the "advantageous" explanation leaves too many unanswered difficulties to sustain a plausible account of the mutual trust that is requisite for human communication.

Comparative psychologist Michael Tomasello's theory of the origins of language is immune to these difficulties, because it provides an integral account of the development of *homo sapiens* as a language user. As a comparative psychologist, Tomasello's point of comparison is human and primate communication. It is important to note here that the comparison between the two species diverges at *gestural* communication, not *vocal* communication. That is to say, primate vocalizations are individual expressions of emotion (not the sort of "conspiratorial whispering" described in Knight's account); by the same token, they do not convey intentionality in the way that physical gestures do. Moreover, primates do not display *cooperative* communication. Like Knight, Tomasello suggests that for humans the development of language and cooperative communication are integral. His theory is grounded in the deceptively simple claim that "we must first understand how humans communicate with one another using natural gestures."[38] Thus, one of Tomasello's significant methodological moves is to look outside of human communication itself and into human cooperation more broadly, especially concepts such as shared intentionality. He makes a convincing case that conventional languages, which are in a certain sense arbitrary, build upon natural gestures, which are not arbitrary.

On this account, communicative gestures in cooperative contexts paved the way for language in modern humans. "Only later still did humans begin to communicate in this new cooperative way outside of cooperative contexts for higher-up, noncooperative purposes—leading to the possibility of deception by lying."[39] Here again we run up against the problem of lying, yet this "problem" supports his case even further. Tomasello notes that lying requires cooperation and an assumed trust among interlocutors. He explains that, while chimpanzees can conceal themselves from others, "there is still no experimental evidence that they can actively mislead others."[40] Lying is a uniquely human phenomenon, and it could only arise in a context of cooperative communication—the kind of communication we find in early humans. Non-human primates "do not communicate cooperatively with the mutual expectation that both participants are attempting to be helpful, including truthful."[41] In human groups where cooperative communication has become the norm, however, lying becomes an alluring prospect. The liar can depend on this expectation of mutual trust and exploit it for selfish purposes.

Thus, in light of Tomasello's reflections, lying is an "unintended consequence" of an otherwise "beautiful tool,"[42] and humans have had to work out a solution to this problem by creating normative sanctions against it. Tomasello (to his credit) does not provide a detailed story of how the problem was overcome, but he suggests that this problem is the *origin* of normative approaches to lying. Here it seems that we have begun to cross the threshold between description and normativity, but I want to resist this move for the time being. Needless to say, I do not share Tomasello's assessment insofar as it suggests that traditional moral sanctions against lying can be reduced to evolutionary expediency. To engage him on this point would take us beyond the scope of this chapter. My primary concern thus far has been to underscore the unique nature of human communication—and of human *deception* in particular. As Tomasello's work has shown us, the problems surrounding the possibility of deception illuminate human language and, by extension, "human nature"—if such a term can be permitted.[43]

INSIGHTS FROM PHILOSOPHY

Finally, I want to consider some recent developments in philosophical ethics and the philosophy of language. First, I turn to the moral philosopher Alasdair MacIntyre, drawing on two strands of his thought. The first strand comes from his *Dependent Rational Animals,* in which he is concerned to reinstate the link between 1) human flourishing and living virtuously and 2) the "metaphysical biology" of human beings.[44] Building on his seminal monograph, *After Virtue,*[45] MacIntyre develops an account of virtue that is based on specific features of human nature, adding the qualifier "dependent" to the classical definition of human beings as rational animals. Thus, the significance of our interdependence plays a central role in MacIntyre's analysis of virtue and social practices.

By the same token, MacIntyre considers the role of language as a central feature of human nature and of human societies. He observes that "language is always embedded in forms of social practice" and that human language is not a result simply of higher biological capacities, but of sociality as an integral feature of human existence.[46] His arguments here dovetail nicely with Tomasello's discussion of joint intentionality, insofar as his understanding of language is grounded in the social animality of human beings.

One of the central foci of *Dependent Rational Animals* is the animality of human nature, and more precisely the observation that human beings, as animals, share certain resemblances to members of other intelligent animal species. "Both initially in our earliest childhood activities and to some significant extent thereafter," he writes, "we comport ourselves to the world in much the same way as other intelligent animals. In transcending some of their limitations we never separate ourselves entirely from what we share with them."[47] In other words, he rejects the picture of human beings that would separate rationality from animality, such that one aspect could be studied independently of the other. It should be noted that, in making this observation, MacIntyre does not mean to erase the distinctiveness of human nature when compared to non-human animals. Rather, he hopes to preserve "the significance of the continuity and resemblance between some aspects of the intelligent activities of non-human animals and the language-informed practical rationality of human beings."[48] Furthermore, "it is of the first importance that what we thereby become are redirected and remade animals and not something else. Our second culturally formed language-using nature is a set of partial, but only partial, transformations of our first animal nature."[49] MacIntyre articulates this insight that our "second" (language-using) nature is a modification of our "first" (animal) nature by pointing to other intelligent, non-human animals and their capacities for thought and feeling. He then argues that having such thoughts or feelings requires language. This is qualified, of course, by certain claims such as that made by MacIntyre that cats, for example, cannot quantify or ascribe modalities to their beliefs.[50] The point, however, is that these intelligent animals are not merely subjects of biological forces—they are "aware," in some sense, of their own relation to their environment. Importantly, MacIntyre also suggests that these intelligent animals are "cultural" insofar as their signs of communication and practices are "conventional" (i.e., they are learned within the context of their own experience and not simply genetic).

This understanding of language as "conventional" informs the second strand of MacIntyre's thought I wish to highlight. In his lectures, *Truthfulness, Lies, and Moral Philosophers*, he offers some important insights that nod toward a kind of "natural normativity" of language. For example, he explains that there is an implicit semantic "rule" in human language prescribing truth-telling—a rule that all of us learn when we learn to speak our native languages: "Note that the rule enjoining truth-telling in speech-acts of assertion is constitutive of language-use as such. It is a rule upon which therefore all interpreters of language-use by others cannot but rely."[51] Even lying presupposes and depends upon this "rule." In order to tell a lie, one must rely upon human regard for truth, even while disregarding it oneself. MacIntyre explains that, according to some moral traditions,

> To tell a lie is wrong as such, just because it is a flouting of the truth, and it is an offense primarily not against those particular others to whom this particular lie has been told, but against human rationality, everyone's rationality, including the liar's own rationality. By lying she or he has failed not only to acknowledge truth as a good that is indispensable in rational relationships with others, but also to recognize that a failure to respect truth is a failure in respecting oneself as a rational being.[52]

This is the point at which concerns about human rationality and concerns about the social dynamics of truth-telling intersect. We recall Aquinas' remark that "It would be impossible for humans to live together, unless they believed one another, as declaring the truth to one another."[53] We are now in a better position to make sense of this claim. Rather than reading this as a simplistic, slippery-slope argument (i.e., "if we don't prohibit lying, then the fabric of society will unravel"), we should understand this as a conceptual claim regarding an undeniable feature of human language, namely, that language use requires an assumption of truth in order to function at all. This is the "semantic rule" that MacIntyre describes, and it does not in and of itself generate any moral claims. In any event, the claim that moral laxity toward lying necessarily poses a threat to human society is somewhat dubious. As MacIntyre notes, "experience goes to show that the social fabric generally survives a good deal" of lying.[54] Yet, on a Thomistic account, the natural goodness of language for human beings will be relevant to the moral goodness of truth-telling, insofar as the will aims at what is good for a person, but a pursuit of this topic would bring us firmly into normative territory. My interest in this brief essay is not to make a normative argument about lying. Rather, my aim has been to draw attention to the fact that these various thinkers have given us a way in which to consider truth-telling as a "natural" function of speech.

CONCLUDING OBSERVATIONS

To conclude, I want to clarify two claims that I do *not* wish to make in this chapter. The first is the claim that empirical observations about human language and its development, in and of themselves, somehow *generate* normative claims about lying. I do not think this, nor did Aquinas. The physical and natural features of our humanity are certainly relevant to our moral considerations, but they cannot be collapsed into simple one-to-one claims about what "ought" to be done. The second is the claim that Aquinas somehow knew or intuited the observations of twenty-first-century anthropologists, psychologists, and moral philosophers when he made his claims about the natural functions of speech. I am not suggesting that these scholars' observations somehow prove that Aquinas was "right" about language.

What I do wish to suggest is that these scholars' observations shed light on Aquinas' arguments, and they provide us with (at least) one model for thinking about truth-telling as an implicit normative feature of language. It is plausible that Aquinas was simply trying to tap into an observable, universal feature of language that we now understand better as a result of contemporary research. If so, then these new observations may actually lend credence to Aquinas' overall analysis of lying, and that, at the very least, this analysis should not be readily dismissed, as it has been in the recent past. Furthermore, I want to suggest that these insights, when read alongside Aquinas, illustrate just how deeply intertwined human language, sociality, rationality, *and* our normative claims about lying (whatever they may be) really are.

Thus, we might begin to consider whether the emergence of human wisdom (a wildly vague term in need of clarification) is coterminous with the emergence of human language. Indeed, we might be tempted to find "wisdom" in the emergence of language in human beings as a species, but this fails to capture the kind of credit or praiseworthiness that we attribute to someone when we use the term "wisdom" colloquially. Can any individual human be considered wise simply because she uses language—a trait that is characteristic of her species as a whole? I suggest, on the contrary, that the phenomenon of lying gives us insight into wisdom by illuminating the ways that humans can manipulate the natural functions of speech. That is (to put it in Thomistic terms), speaking truthfully is an act of the will, and the virtuous and wise language user will learn *when* to speak the truth and in *what manner*. Language is natural for humans, but lying and truth-telling take us beyond the natural and into the normative. Yet, fascinatingly, contemporary research has revealed that lying and truth-telling are constitutive of the development of language itself, even if we endeavor to set aside their normative implications. Perhaps these insights suggest that the dividing line between "natural" and "normative" is not as impermeable as it is often assumed to be.

What contemporary researchers help us understand is that language is neither "natural" in the sense that it is a direct byproduct of human biology, nor is it an "artifact" in the sense that it is arbitrary or lacking any inherent normative constraints. It is a product of evolution and human social cooperation, but it is nonetheless fundamentally *human*. On

Aquinas' understanding of morality, to flourish as a human being is to flourish as a language speaker. This does not mean that a person must exhibit a mastery of vocabulary and syntax in order to be virtuous—although such a mastery would indeed reflect certain intellectual virtues—but it does mean that the morally virtuous person will necessarily exercise her virtue in the way she uses words to communicate with other human beings. Language is not an accidental feature of human existence or an optional appendix to human flourishing. Moreover, if Tollefsen is right that language is an artifact, then we can see how Aquinas does indeed fall prey to the Is-Ought fallacy. If my analysis is sound, however, then Tollefsen cannot claim that language is merely an artifact. And if language really is natural, then we cannot simply dismiss Aquinas' claim about the natural function of speech. Aquinas' argument may still fall prey to the so-called "Naturalistic Fallacy," but that is not an argument I can engage in this chapter.

We must keep in mind that Aquinas never claims that language has only one function. Nor does he claim that the natural function(s) of language automatically generate(s) normative moral claims. Rather, we should understand the argument as fundamentally concerned with the habituation of moral virtue—in this case, the virtue of truth. The additional argumentation and the "missing premises," so to speak, are to be found in his voluminous writings on the virtues. The point is that the cultivation of virtue requires that we develop certain character traits grounded in our human capacities and natural inclinations. Aquinas offers a multifaceted approach to the virtue of truth, demonstrating along the way its inherent relationship with the cardinal virtue of justice. In addition to this theoretical framework, it is equally important that his claims about truthfulness (and by extension, his opposition to lying) are grounded in observable features of human existence. This is where the "natural function" of speech becomes relevant. Contemporary scholars have shown that there is a sense in which truthfulness is normative in human language by its very nature, although not in the robust, moral sense at which Aquinas' argument ultimately arrives. But what this contemporary research does confirm is that truth is integral to language, at a level far deeper than even Aquinas himself could have perceived.

NOTES

1. The literature on this topic is plentiful. For some representative examples of cross-cultural studies on lying, see Gail D. Heyman et al., "Instrumental Lying by Parents in the US and China," *International Journal of Psychology* 48.6 (2013): 1176–84 and Susan D. Blum, *Lies That Bind: Chinese Truths, Other Truths* (Lanham, MD: Rowman and Littlefield, 2007).

2. Angela D. Evans and Kang Lee, "Emergence of Lying in Very Young Children," *Developmental Psychology* 49.10 (2013): 1958–63. See also Michael Lewis, "The Origins of Lying and Deception in Everyday Life," *American Scientist* 103.2 (2015): 128–35.

3. Paul J. Griffiths, *Lying: An Augustinian Account of Duplicity* (Grand Rapids, MI: Brazos, 2004), 11.

4. Comparative psychologist Michael Tomasello notes, "Interestingly and importantly, whereas chimpanzees can conceal themselves from others . . . there is still no experimental evidence that they can actively mislead others—perhaps because, to repeat, lying requires cooperative communication." Michael Tomasello, *Origins of Human Communication* (Cambridge, MA: MIT Press, 2008), 190.

5. See, for example, Thomas Aquinas, *Summa Theologiae,* trans. Laurence Shapcote, O.P., ed. John Mortensen and Enrique Alarcón, 8 vols. (Lander, WY: The Aquinas Institute for the Study of Sacred Doctrine, 2012), 2a2ae 110.3, which I discuss below.

6. Augustine, *On Christian Teaching* [*De Doctrina Christiana*], trans. R. P. H. Green (New York: Oxford University Press, 2008), II.99.

7. For instance, Melvin Konner writes, "Language . . . may have as one of its major functions deception—the covering up of base motives or the distortion of them to make them look pure. This, indeed, may be one of the major adaptive purposes of communication systems in many animals, in an evolutionary sense. And we may have carried it to its most exotic realization—a communication system so complex and flexible that it raises to utter perfection the self-serving enterprise of lying." Melvin Konner, *The Tangled Wing: Biological Constraints on the Human Spirit* (New York: Holt, Rinehart, and Winston, 1982), 169.

8. Aquinas, *Summa Theologiae* 2a2ae 110.3: "Est enim actus cadens super indebitam materiam, cum enim *voces sint signa naturaliter intellectuum,* innaturale est et indebitum quod aliquis voce significet id quod non habet in mente" (italics added to indicate the translated portion). Latin citations of the *Summa* are from the Leonine edition found in Aquinas, *Summa Theologiae,* ed. Mortensen and Alarcón. All English translations are my own. Aquinas refers to words as "signa intellectuum" twenty-two times throughout his works, but, curiously, this is the only instance in which he adds the qualifier "naturaliter." A more literal translation might read, "Spoken words are naturally signs of the intellect," where *intellectuum* is Aquinas' standard term for the higher cognitive faculty. Given the context of this passage, however, he is simply trying to convey the notion that words are verbal signifiers of one's thoughts.

9. Aquinas is drawing here on Aristotle's view that words are signs of one's thoughts, although Aristotle specifically states that spoken words are signs of the "passions of the soul": *Ἔστι μὲν οὖν τὰ ἐν τῇ φωνῇ τῶν ἐν τῇ ψυχῇ παθημάτων σύμβολα* (*Peri Hermeneias* I.1). It is worth noting that Aristotle does not deploy this principle to argue against lying. Rather, Aristotle's objections to lying (*Nicomachean Ethics* IV.7) are framed within a spectrum that includes boasting and false humility, both of which extend beyond the mean of virtue. Thus, Aquinas, in typical fashion, is drawing upon Aristotle in order to go beyond what the Philosopher says himself. For an insightful analysis of Aristotle's position on lying, see Jane S. Zembaty, "Aristotle on Lying," *Journal of the History of Philosophy* 31.1 (1993): 7–29. Zembaty concludes (correctly, in my judgment) that, while Aristotle holds a generally negative view of lying, for him some lies are morally acceptable "when they harm no one, involve no undeserved disrespect, and stem from excellence and self-sufficiency rather than some deficiency of character" (29).

10. Christopher O. Tollefsen, *Lying and Christian Ethics* (Cambridge: Cambridge University Press, 2014), 45.

11. The *locus classicus* of the "Is-Ought" fallacy can be found in David Hume, *A Treatise Concerning Human Nature,* ed. David Fate Norton and Mary J. Norton (New York: Oxford University Press, 2000), 3.1.1. Hume observes that moral philosophers, when describing a state of affairs (the way things are), often slip imperceptibly into normative statements: "For as this *ought,* or *ought not,* expresses some new relation or affirmation, 'tis necessary that it shou'd be observ'd and explain'd; and at the same time that a reason should be given, for what seems altogether inconceivable, how this new relation can be a deduction from others, which are entirely different from it." Hume scholars in fact debate the correct reading of this text, which has been inter-

preted variously to support non-cognitivsm, to refute naturalistic moral arguments, or simply to criticize what Hume considered intellectual laziness. Nonetheless, Hume became associated with the so-called "Is-Ought" fallacy, which now functions as a sort of lightning rod among moral philosophers.

12. Tollefsen, *Lying and Christian Ethics*, 45.

13. Ibid., 47.

14. Aquinas, *Summa Theologiae*, 2a2ae 109.2: "Est autem specialis quidam ordo secundum quod exteriora nostra vel verba vel facta debite ordinantur ad aliquid sicut signum ad signatum. Et ad hoc perficitur homo per virtutem veritatis."

15. Aquinas, *Summa Theologiae*, 2a2ae 58.1: "[I]ustitia est habitus secundum quem aliquis constanti et perpetua voluntate ius suum unicuique tribuit."

16. For a recent and lucid Thomistic analysis of the virtue of justice, particularly on the intersection of virtue and moral precepts, see Jean Porter, *Justice as a Virtue: A Thomistic Perspective* (Grand Rapids, MI: Eerdmans, 2016), 38–42 and 151–52.

17. Aquinas, *Summa Theologiae*, 2a2ae 109.3.

18. Aquinas, *Summa Theologiae*, 2a2ae 110.3.

19. Aquinas, *Summa Theologiae*, 2a2ae 109.3 ad 1.

20. Chris Knight offers a more radical view, in which speech is "a revolutionary development made possible by the establishment of novel levels of social co-operation." Chris Knight, "Ritual/Speech Coevolution: A Solution to the Problem of Deception," in *Approaches to the Evolution of Language: Social and Cognitive Bases*, ed. James R. Hurford, Michael Studdert-Kennedy, and Chris Knight (New York: Cambridge University Press, 1998), 68. But the more basic notion that the development of language requires mutual trust has become commonplace in the literature (see references to Tomasello below).

21. Knight cites the work of Amotz Zahavi, who "concludes that, since potential conflicts of interest exist throughout the animal world, even between close kin, resistance to deception has always selected against conventional signals—with the one puzzling exception of humans." Knight, "Ritual/Speech Coevolution," 72.

22. Displaced reference is "the ability to refer to an item that has been moved (displaced) in space and/or time." See H. Lyn, J. L. Russell, D. A. Leavens, et al., "Apes Communicate about Absent and Displaced Objects: Methodology Matters," *Animal Cognition* 17 (2014): 85. As the authors note, displaced reference "has been called one of the true hallmarks of referential communication." This article in particular notes the methodological difficulties in determining whether non-human primates possess this ability, citing the conflicting results of several studies.

23. Michael Arbib has recently offered one of the most thorough and sophisticated counterhypotheses to this model with his "Mirror System Hypothesis." As he explains, "the *H. sapiens* brain was *language ready*, but . . . it took tens of millennia of cultural evolution before human had languages and *language-using brains*, even though the brain genome had not changed significantly in the interim. More specifically, the hypothesis is that biological evolution and cultural evolution had together yielded brains and social structures in early *H. sapiens* that could support rudimentary systems of communication, *protolanguages*, with a somewhat open set of "protowords," but not *languages* in the sense of communication systems endowed with rich and open-ended lexicons and grammars supporting a compositional semantics. It thus took cultural evolution—*niche construction* . . .—to yield societies in which *language-ready* brains could become *language-using* brains." Michael Arbib, "Toward the Language-Ready Brain: Biological Evolution and Primate Comparisons," *Psychonomic Bulletin and Review* 24 (2017): 142.

24. Chomsky has since developed his views in many books and articles, but the seminal monograph was *Aspects of the Theory of Syntax* (Cambridge, MA: MIT Press, 1965).

25. For an accessible summary of the reasons why Chomsky's understanding of language is now being overturned, see Paul Ibbotson and Michael Tomasello, "Language in a New Key," *Scientific American* 315.5 (2016): 70–75.

26. In Dunbar's later writings he cites Knight approvingly, but he is slightly skeptical of the claim that language evolved specifically to make ritual possible (see below). Dunbar's own view is that language "evolved to facilitate bonding through the exchange of social information, and was later hijacked for use in . . . semi-religious contexts in such a way as to formalize what may well have been preexisting practices." Robin Dunbar, *Grooming, Gossip, and the Evolution of Language* (Cambridge, MA: Harvard University Press, 1996), 147.

27. Knight, "Ritual/Speech Coevolution," 81–87.

28. Ibid., 74.

29. Ibid., 72.

30. "Displaced reference" is defined above (n. 21). "Duality of patterning" here refers to the ability to make meaning out of speech on two distinct levels: to combine sounds (e.g., syllables) into words, and to combine words into meaningful statements (e.g., propositions).

31. Ibid., 82.

32. Ibid., 83–87.

33. Terrence W. Deacon, *The Symbolic Species: The Co-Evolution of Language and the Brain* (New York: Norton, 1997), 409–10.

34. Ibid., 376–79.

35. Ibid., 410.

36. See especially the work of Michael Arbib (cited above) and Thomas Scott-Phillips (cited below).

37. Jean-Louis Dessalles, "Why Talk?" in *The Social Origins of Language* , ed. Danny Dor, Chris Knight, and Jerome Lewis (New York: Oxford University Press, 2014), 284–96.

38. Tomasello, *Origins*, 2.

39. Ibid., 8.

40. Ibid., 190.

41. Ibid., 217. For a more recent study on primate deception (with which Tomasello was involved), see Katja Karg et al., "Chimpanzees Strategically Manipulate What Others Can See," *Animal Cognition* 18 (2015): 1069–76.

42. Tomasello, *Origins*, 190.

43. Anthropologist Thomas Scott-Phillips has recently leveled some criticisms against some of the common assumptions and methodologies in comparative psychology, including Tomasello's. He agrees with Tomasello, however, that human ostensive communication laid the evolutionary groundwork for linguistic communication. He argues that in fact not enough attention has been paid to ostension (i.e., assertion by means of gesture or symbol; this is the counterpart of *inference*) as an underlying component of language and that even less attention has been paid to the role of pragmatics (which is typically the domain of philosophers of language). His arguments fill some of the lacunae in the relevant literature in anthropology and developmental psychology, but as far as I can tell, they do not undermine the central theses of Tomasello's *Origins of Human Communication*, particularly regarding the role of deception in human evolution, which is my primary concern in this chapter. See Thomas Scott-Phillips, "Nonhuman Primate Communication, Pragmatics, and the Origins of Language," *Current Anthropology* 56.1 (2015): 56–80.

44. MacIntyre explains his late embrace of metaphysical biology as follows: "Although there is indeed good reason to repudiate important elements in Aristotle's biology, I now judge that I was in error in supposing an ethics independent of biology to be possible—and I am grateful to those critics who argued this case against me—and this for two distinct, but related reasons. The first is that no account of the goods, rules and virtues that are definitive of our moral life can be adequate that does not explain—or at least point us towards an explanation—how that form of life is possible for beings who are biologically constituted as we are, by providing us with an account of our development towards and into that form of life. That development has as its

starting point our initial animal condition." Alasdair MacIntyre, *Dependent Rational Animals: Why Human Beings Need the Virtues* (Chicago: Open Court, 1999), x.

45. Alasdair MacIntyre, *After Virtue: A Study in Moral Theory*, 3rd ed. (Notre Dame, IN: University of Notre Dame Press, 2007).

46. MacIntyre, *Dependent Rational Animals*, 34.

47. Ibid., 8.

48. Ibid., 50.

49. Ibid., 49.

50. Ibid., 38.

51. Alasdair MacIntyre, *Truthfulness, Lies, and Moral Philosophers: What Can We Learn from Mill and Kant?* (The Tanner Lectures on Human Values, 1994), 312. http://tannerlectures.utah.edu/_documents/a-to-z/m/macintyre_1994.pdf.

52. Ibid., 315.

53. Aquinas, *Summa Theologiae*, 2a2ae 109.3 ad 1.

54. MacIntyre, *Truthfulness, Lies, and Moral Philosophers*, 328.

BIBLIOGRAPHY

Aquinas, Thomas. *Summa Theologiae*. Translated by Laurence Shapcote, O.P. Edited by John Mortensen and Enrique Alarcón. 8 vols. Lander, WY: The Aquinas Institute for the Study of Sacred Doctrine, 2012.

Arbib, Michael. "Toward the Language-Ready Brain: Biological Evolution and Primate Comparisons." *Psychonomic Bulletin and Review* 24 (2017): 142–50.

Aristotle. *Categories, On Interpretation, Prior Analytics*. Translated by H. P. Cook and Hugh Tredennick. Loeb Classical Library. Cambridge, MA: Harvard University Press, 1938.

———. *Nicomachean Ethics*. Translated by H. Rackham. Loeb Classical Library. Cambridge, MA: Harvard University Press, 1934.

Augustine. *On Christian Teaching*. Translated by R. P. H. Green. New York: Oxford University Press, 2008.

Blum, Susan D. *Lies That Bind: Chinese Truths, Other Truths*. Lanham, MD: Rowman and Littlefield, 2007.

Chomsky, Noam. *Aspects of the Theory of Syntax*. Cambridge, MA: MIT Press, 1965.

Deacon, Terrence W. *The Symbolic Species: The Co-Evolution of Language and the Brain*. New York: W. W. Norton, 1997.

Dessalles, Jean-Louis. "Why Talk?" In *The Social Origins of Language*, edited by Danny Dor, Chris Knight, and Jerome Lewis, 284–96. New York: Oxford University Press, 2014.

Dunbar, Robin. *Grooming, Gossip, and the Evolution of Language*. Cambridge, MA: Harvard University Press, 1996.

Evans, Angela D., and Kang Lee. "Emergence of Lying in Very Young Children." *Developmental Psychology* 49.10 (2013): 1958–63.

Griffiths, Paul J. *Lying: An Augustinian Account of Duplicity*. Grand Rapids, MI: Brazos, 2004.

Heyman, Gail D., Anna S. Hsu, Genyue Fu, and Kang Lee. "Instrumental Lying by Parents in the US and China." *International Journal of Psychology* 48.6 (2013): 1176–84.

Hume, David. *A Treatise Concerning Human Nature*. Edited by David Fate Norton and Mary J. Norton. New York: Oxford University Press, 2000.

Ibbotson, Paul, and Michael Tomasello. "Language in a New Key." *Scientific American* 315.5 (2016): 70–75.

Karg, K., M. Schmelz, J. Call, and M. Tomasello. "Chimpanzees Strategically Manipulate What Others Can See." *Animal Cognition* 18 (2015): 1069–76.

Knight, Chris. "Ritual/Speech Coevolution: A Solution to the Problem of Deception." In *Approaches to the Evolution of Language: Social and Cognitive Bases*, edited by James

R. Hurford, Michael Studdert-Kennedy, and Chris Knight, 68–91. New York: Cambridge University Press, 1998.

Konner, Melvin. *The Tangled Wing: Biological Constraints on the Human Spirit*. New York: Holt, Rinehart, and Winston, 1982.

Lewis, Michael. "The Origins of Lying and Deception in Everyday Life." *American Scientist* 103.2 (2015): 128–35.

Lyn, H., J. L. Russell, D. A. Leavens, K. A. Bard, S. T. Boysen, J. A. Schaeffer, and W. D. Hopkins. "Apes Communicate about Absent and Displaced Objects: Methodology Matters." *Animal Cognition* 17 (2014): 85–94.

MacIntyre, Alasdair. *After Virtue: A Study in Moral Theory*, 3rd edition. Notre Dame, IN: University of Notre Dame Press, 2007.

———. *Dependent Rational Animals: Why Human Beings Need the Virtues*. Chicago: Open Court, 1999.

———. *Truthfulness, Lies, and Moral Philosophers: What Can We Learn from Mill and Kant?* The Tanner Lectures on Human Values, 1994. Available online: http://tannerlectures.utah.edu/_documents/a-to-z/m/macintyre_1994.pdf.

Porter, Jean. *Justice as a Virtue: A Thomistic Perspective*. Grand Rapids, MI: Eerdmans, 2016.

Scott-Phillips, Thomas. "Nonhuman Primate Communication, Pragmatics, and the Origins of Language." *Current Anthropology* 56.1 (2015): 56–80.

Tollefsen, Christopher O. *Lying and Christian Ethics*. Cambridge: Cambridge University Press, 2014.

Tomasello, Michael. *Origins of Human Communication*. Cambridge, MA: MIT Press, 2008.

Zembaty, Jane S. "Aristotle on Lying." *Journal of the History of Philosophy* 31.1 (1993): 7–29.

SEVEN

Precursors to Explanations of Action

Collective Intentionality and the Wisdom of Early Childhood

Craig Iffland

In this chapter, I examine a particular social practice, namely, the practice of offering explanations of action. My interest in exploring this practice is that moral discourse presupposes not only the intelligibility of asking for and giving explanations of action, but also the capacity to inhabit a normative, social order that provides common standards for the rectitude of such explanations. In what follows, I focus on the capacity for *collective intentionality*, cognition of how we—as members of a social group—collectively relate to some particular, common object of attention.[1] This capacity enables young children to acquire the art of navigating a social world that routinely demands that they explain themselves. A key manifestation of this capacity is young children's ability to forge and comprehend a shared, experiential background or social frame with interlocutors. By relating to particular objects of cognition within this shared frame, normative standards of evaluation and explanation of action become possible.

In the first section, Critiquing Theory of Mind, I begin by setting out some disagreements with the way some scholars have portrayed the comprehension of mental states by young children, focusing critical attention on the claim that possession of a "theory of mind" is an important achievement in cognitive development. In the second section, Collective Intentionality and Normative Explanations of Action, I offer empirical support for an alternative view of human cognition grounded in our capacity for collective intentionality, paying particular attention to joint attention as a manifestation of that capacity in children from early infancy onwards. A key claim advanced is that joint attention allows children to forge a common perspective on the world with others, a way that they, together with others, collectively engage the world. Such a perspective is already normative in some sense, enabling children to make sensible explanations of particular actions. Finally, in the section entitled Thomas Aquinas and the Evolution of Practical Wisdom, I conclude by reflecting on the significance of these studies for our understanding of the evolution of practical wisdom in light of some of Thomas Aquinas's reflections on the virtue of prudence.

CRITIQUING THEORY OF MIND

A cursory perusal of the literature on the cognitive capacities of infants and young children reveals a field fascinated with the study of the emergence of a child's *theory of mind,* which, so it is supposed, serves as a cornerstone in cognitive development. Oddly enough, theory of mind, although a concept deployed for a wide array of research on human and non-human cognition, originates from studies of primate cognition and marks an attempt to explain the extent to which chimps can infer the mental states of other agents based on observation of the behavior of others. David Premack and Guy Woodruff, who originally coined the term, describe "theory of mind" in the following manner:

> In saying that an individual has a theory of mind, we mean that the individual imputes mental states to himself and others . . . A system of inferences of this kind is properly viewed as a theory, first because such states are not directly observable, and second because the system can be used to make predictions, specifically about the behavior of other organisms.[2]

The definition is not innocent of contestable assumptions. The definition assumes that our concepts of intention, knowledge, and belief signify mental states that are *not* directly observable. Because these mental states are not directly observable, our use of these concepts is based on an inference from an organism's *behavior* that it is in a particular mental state. With this inference in tow, one can thereby adequately *predict* the organism's behavior. The underlying cognitive system or processes that enable this capacity for inference and prediction is identified as the organism's theory of mind.

Let me begin with the assertion that mental states are not directly observable, but rather inferred on the basis of what is directly observable, i.e., another's behavior. This dichotomy between the observable (behavior) and the unobservable (mental states) has a history. The history unfolds in terms of two revolutions in psychology: the earlier turn to behaviorism and the later turn to cognitivism. A central contention of the former revolution was that mental events, unlike behaviors, are not "publicly observable" and so lack scientific objectivity.[3] As George Miller explains, the former revolution "proposed to redefine psychology as the science of behavior" because "by changing the subject to the study of behavior, psychology could become an objective science based on scientific laws of behavior."[4] According to one of the central proponents of this revolution, John Watson, a behaviorist views psychology as a "purely objective experimental branch of natural science" whose aim is "to predict, given the stimulus, what reaction will take place; or, given the reaction, state what the situation or stimulus is that has caused the reaction."[5] As Alan Costall and Ivan Leudar observe, the mid-twentieth-century return to cognitivism is often conceived as a revolt against such behaviorist theories even though many cognitive psychologists retained some version of methodological behaviorism.[6] Thus, a recent textbook on cognitive psychology claims that

> The modern study of cognition is concerned with mental processes such as perceiving, remembering, reasoning, deciding, and problem solving . . . it assumes (1) only by studying mental processes can we fully understand what organisms do, and (2) we can study mental processes in an objective fashion by focusing on specific behaviors (just as behaviorists do) but interpreting them in terms of underlying mental processes.[7]

The above quote perfectly illustrates the methodological presuppositions at work in present-day theory of mind research. As with the behaviorists, there remains an explanatory gap between behavior and mental processes, but contrary to the behaviorists, this gap can be bridged by an interpretation of behavior by appeal to some mental state. To bridge this gap is to possess a theory of mind, a complex system of inferences from behavior to mental states (and vice versa) that enables a level of predictive capacity in the agent who possesses it. By describing cognition and cognitive development in terms of a theory of mind, however, one conveniently re-inscribes Watson's behaviorist scientific program of prediction and control into the nature of cognition. Doing so transforms infants and primates into "proto-scientists, who treat other people with detachment, objectively, and in an intellectual mode" and assumes further that intrinsic to the cognitive processes of humans and other primates is the dichotomy between an agent's observable behavior and her invisible mind.[8] Can this be right?

If mental states are not directly observable, but only inferred on the basis of another's behavior, it follows that we do not directly observe purposive or intentional action. Imagine yourself witnessing a man chopping wood in his backyard. "Chopping" cannot be something directly observable to you since it is a description of what the man is doing that appeals to a psychological state, i.e., an intention.[9] Nor can you directly observe the movement of the man's arms or his axe "up" or "down," since such a description presupposes an apprehension that the man is doing something on purpose, intentionally with his arms or axe. All we can directly observe of the man is a variety of disconnected bodily positions—a series of snapshots of the physical motions involved in his chopping wood. According to the theory of mind advocate, from this series of snapshots one can infer an intention on the part of the man. Given this inference to a particular mental state, we can formulate a hypothesis about his future behavior. The question is how one can make any valid inference from observation of non-psychological behavior to a mental state of intention? If one cannot observe intentions directly, our series of snapshots of physical motions cannot be put into any kind of teleological order—it must be a random series, not a picture show, if it is to respect the claim that we do not observe intentions. Could one plausibly infer from a series of snapshots that a man is chopping wood intentionally?

Suppose that we took all the snapshots at random. If one is entitled to observe the "axe," it is plausible that we could infer that a man is chopping wood. But one is not so entitled. For an axe cannot disclose its purpose to us any more than a man's behavior does. In order to observe something as an axe, one must already have in one's possession a concept of the *purpose* of this particular cultural artifact, namely, that an axe is something used to cut things. Now, suppose an individual possesses the concept "axe." How did she come to acquire the concept? She learned it of course! When she was younger she asked her father, "what is Paul Bunyan carrying?" and he answered, "an axe." Yet, as John O'Callaghan has argued, this move presupposes that the daughter interprets the behavior of the father—the random movement of his lips and the random noises conjoined to this behavior—as a speech act, an intention to communicate.[10] Because she has a theory of mind, this interpretation is probabilistic, a best guess at interpreting the behavior. But how does she get the notion of speech as a kind of action that she guesses her father intends to perform when all she ever *directly* encounters and observes is non-psychological behavior rather than intentional action? In other words, we are not entitled to assume that our subject could see an "axe" or acquire concepts through communication with others without already postulating that the gap between observable behavior and invisible mind has been crossed. I do not think it is possible to cross this chasm. As Leudar and Costall put it, "there is *no* way to bridge the gap between what we can experience of others and an understanding of their intentions, emotions and so on, given that the only grounds for such understanding is supposed to be behavior *divested of all psychological meaning*."[11] In this respect, theory of mind requires children to do the impossible. How is it then that children routinely evidence a "theory of mind?"

One reason for their success is that theory of mind researchers are not as committed to methodological behaviorism as the canonical definition of the theory suggests. Researchers work with a highly restricted notion of a mental state, and as a consequence, routinely treat descriptions of the intentionality of behavior as the basis for a child's inference to another's mental state of intending. For instance, one group of researchers asked how children select from different "knowledge systems" in order to provide psychological, physical, or biological explanations of some motion or event.[12] The researchers opined that

> [O]ne possibility is that children apply these reasoning systems in accord with a related analysis of the properties of the event to be explained: if the motion is intentional, if it is goal-directed, if the situation is psychological in some larger sense, then explain it in terms of psychological constructs regardless of the entities involved [in the situation].[13]

Here we must ask the following: if possession of a theory of mind means a system of inferences from observable behavior to unobservable mental states, how can children directly observe the intentional or goal-directed character of an event, a motion, or a situation *in order to* select the appropriate knowledge system to be applied? The researchers assume the very capacity—knowledge of the intentional character of motions, events, and situations—that the studies purport to demonstrate. Given our analysis regarding the problems of applying a strict division between behavior and mind, this should come as no surprise. The system of inference metaphor only works if we *assume* the intentional character of the behavior from the outset. In other words, it is because we *directly observe* that the man is chopping word that we can then infer that he *wants* to chop the wood; it is because we *directly observe* the man raising his axe above some wood that we can infer that he *intends* to "chop the wood." At this point, however, we must ask whether inference is the best description of what is happening in our judgment that the man "intends" to chop wood.

Only an alien from another planet would need to infer what the man intends to do in these circumstances. When we see a man with an axe, outdoors, setting up wood on a stand, we are entitled to say, "Ah, he intends to chop some wood." The claim evidences not some attribution of a mental state, but an awareness of the convention of wood chopping. It certainly is not an "educated guess." Take another example drawn from a more familiar social practice. Within the context of a penalty kick, one does not infer that the player will strike the ball with her foot in the direction of the goal; rather, one *knows* that she will do so because one *knows* how *we* play "soccer" or "football." Of course, someone completely unfamiliar with the game may only see a woman, a ball, a wide net, a woman in front of the net, a crowd, etc., and infer that this woman *may* kick the ball in the direction of the wide net. Such an inference is made against a backdrop of ignorance about the larger context in which this event is situated. For the most part, neither adults nor children operate against such a backdrop because, by and large, they move with relative ease within the larger socio-cultural context of which they are a part. For this reason, I would argue that failure to explain *why* one does something indicates paucity of socio-cultural understanding rather than a failure to *construct* or *access* a system of inferences from behavior to mind.

Consider one such failure noted by the authors of the study quoted from above. A child is told a story that is supposed to disclose a mistaken belief. In the story, a character (Sarah) wants to have ice cream with chocolate sauce for a snack. Taking a "squeeze bottle" out of the refrigerator, she squeezes it onto her ice cream. At this point, the adult tells the child, "look, Sarah squeezes ketchup on her ice cream. Why did that happen?" Some children catch that this was a mistake and explain it as such, i.e., "she just didn't know it was ketchup." Others will say, "because she just wanted to try it on her ice cream." The latter are assumed to give the wrong kind of psychological explanation because they have not connected the character's "action" (squeezing ketchup on her ice cream) to her false belief that "the squeeze bottle she took contained chocolate sauce." What best accounts for the purported failure—a deficiency in a cognitive system of inferences or socio-cultural understanding?

The story presupposes an understanding of certain social conventions. Ice cream is a "snack," chocolate sauce is a conventional topping for ice cream, ketchup is an inappropriate topping for ice cream, and (less credibly) the similarities between a ketchup and a chocolate sauce "squeeze bottle" are sufficient to make it reasonable for someone to mistake a ketchup bottle for a bottle of chocolate sauce. In order for a child to describe Sarah's action in terms of a false belief, she has to make all of these connections. This is a function of her *level of socialization*, however, not her cognitive processing skills. It is not uncharacteristic for young children to want to try a combination of foods that an adult would reject by appeal to some conventional standard. At a certain point, however, a child learns that "ketchup" goes with hamburgers and "chocolate sauce" goes with ice cream. This is a cognitive advancement, of course, but only in the sense that it marks an expansion of the child's apprehension of the socio-cultural standards that govern how *we* eat things. A mistake is intelligible only as the breach of some standard *we* follow. In other words, because *we* put chocolate sauce on ice cream, and because *we* do not put ketchup on ice cream, putting ketchup on ice cream is intelligible as a mistake. What one tests in such stories is not the development of the child's theory of mind, but her capacity to make judgments in light of *our* standards of eating.

So, attribution of a false belief to some agent presupposes an awareness of a socio-cultural standard to which the agent's action fails to conform. The criterion for the child's judgment that Sarah did not intentionally pour the ketchup on the ice cream is *normative*, drawing upon a shared social norm, rather than *inferential*, making an inference about a mental state based on observation of Sarah's behavior. One might say, "since we *do not* pour ketchup on ice cream, Sarah must be pouring ketchup on ice cream *by mistake*; she must have thought it was chocolate sauce." With this connection made, we look around for an alternative explanation of the action, "ah, she must have confused the bottles." What is on display here is the child's competency in navigating a particular "form of life" shared with the researchers rather than a child's deployment of a theory of mind.[14] In fact, as these studies indicate, young children do not need to observe any behavior—but can merely listen to stories—in order to apprehend the intentional character of an agent's action. This is because the standards governing our use of this psychological term are dependent upon our inhabiting a shared experiential frame—a narrative or story—that determines the intelligibility of their application.

What has been argued thus far? I began by noting that the theory of mind framework presupposes methodological behaviorism while nonetheless trying to explain non-psychological behavior by appeal to hidden cognitive processes. I offered reasons why it is impossible to justify any inference from non-psychological behavior to mind without already assuming intentional elements in the behavior or the larger context in which that "behavior" is situated. Thereafter, I argued that our psychological explanation of action is *normative* rather than *inferential*. To see a "mistaken" action is to see it *as* a violation of some socio-cultural standard of expectation. The identification of an agent's intentions as well as of her mistakes depends upon our viewing the agent against the backdrop of our knowledge of these standards—what I identified as the experiential frame. The real cognitive achievement in children is not the "detached observation and predictive" mode of cognition championed by theory of mind advocates, but rather a normatively-loaded, socially mediated mode of cognition that enables children to enter into a shared experiential frame with their interlocutors.

In making this claim, I align myself with a particular account of social cognition in human beings—namely, their capacity for collective intentionality, to cognize in a collective or we mode. In the following section, I underline the disagreement between this model of social cognition and the one provided by theory of mind. Thereafter, I provide some empirical support for collective intentionality in young children that helps to explain the capacity of young children to enter into the practice of requesting and offering explanations of action.

COLLECTIVE INTENTIONALITY AND NORMATIVE EXPLANATIONS OF ACTION

Considered as foundational models of social cognition, the contrast between theory of mind and collective intentionality could not be more stark. While the theory of mind account of social cognition presupposes an individualistic, detached observer—a "proto-scientist"—advocates of collective intentionality emphasize the extent to which our cognitive skills are shaped by our continuing interactions with others. The former, I think, presupposes a level of abstraction and detachment from the social, cultural, and conversational context in which *human beings* regularly explain things (and themselves) to one another.[15] As Mattia Galloti and Chris Frith helpfully explain,

> The spectatorial view of social cognition accounts for the behaviour of interacting agents from the perspective of an observer *qua* theorist, who represents the decision problem faced by the agents as it appears to him. Yet, engaging in interaction changes the way in which interacting agents understand the problem, in that contextual features of the interactive scene prime representations that are not available to the isolated "theorist."[16]

As any anthropologist could attest, what is opaque to a detached observer of some practice can be utterly obvious to one of its participants. It is precisely the acquisition of the latter perspective that enables one to explain the activity with relative ease. By continued interaction with participants, the theorist comes to see not only what is and what is not relevant to the explanation of some practice, but also the larger, social context in which the practice takes place. Where the theory of mind advocate treats the developing child as an emerging natural scientist, advocates of collective intentionality treat her as an emerging anthropologist. Cognitive advancements are a function of the child's progressive learning and adoption of a particular form of life—its norms, customs, and language.[17] These advancements are made possible through her interaction with others, enabling her to adopt a collective experiential frame through which various practices and actions become intelligible. As a participant in a history of social interactions of this kind, she is more able to ascertain what "we" think is relevant for the explanation of some particular action. Explanation is an outgrowth of social interaction and conversation rather than an innate theory of mind.

This is why we can more easily understand Sarah's pouring ketchup on her ice cream as *intentional* if this is seen against the backdrop of a prior interaction with Sarah in which she expressed a predilection for ketchup as an ice cream topping. Such interaction need not be person-to-person, however. Absent some prior interaction in which Sarah expresses her distaste for ketchup on ice cream, the seemingly natural intuition that

Sarah's pouring of ketchup on her ice cream is a *mistake* depends upon our seeing the action against the backdrop of a host of other interactions. It is through these interactions that the child acquires an understanding of how *we* use ketchup or *our* standards for ice cream toppings. The capacity of a child to apprehend Sarah's mistake—"she didn't know it was ketchup"—is a function of her capacity for collective intentionality, to cognize from a collective, social frame that renders the action intelligible to *us* as a mistake.[18]

This does not mean that the child explicitly adverts to this frame in describing Sarah's action as a mistake, but only that her capacity to give such a description presupposes some kind of habitual knowledge of certain social conventions that provides the frame in light of which Sarah's action can be understood as a mistake. So, while it is true that when children apprehend intentions, mistakes, or false beliefs, they demonstrate a particular cognitive skill, what is important is that this skill is "shaped by . . . even created by, their interactions with others in the culture or with the artefacts and symbols that others have created for communal use."[19] The idea that cognitive skills, at least in human beings, are intrinsically social and normative lies at the heart of those proposals that identify the capacity for collective intentionality as a unique feature of human cognition. How does this understanding of social cognition help us explain the cognitive underpinnings of the social practice of giving explanations of action?

One answer is the capacity for joint attention—i.e., the intentional, mutual engagement with some common object—which, in human beings, is an early manifestation of collective intentionality.[20] Mutual engagement with an object depends not only on a convergence of motive, but also on the ability of one party to call attention to the object. If we are to have a meaningful agreement or disagreement about the explanation of an action, we must jointly attend to it. One may ask, "why is he lying down?" but this is not the only way one draws attention to an action. I may *point* to an action that calls for explanation without any accompanying verbalization. In doing so, I attempt to forge a shared perspective from which to view the "thing." For "through pointing, I identify something for us to attend to. My pointing thus presents us with a topic, an object of predication."[21] By pointing at some "thing," I enable *us* to make statements about "what" or "how" it is—whether it is an action (killing) or a color (red).

In human beings, the use of pointing comes about at a very early age. Infants as young as eleven months will (a) look in the direction indicated by a pointing finger,[22] (b) point out objects of interest for adults based on prior interactions with them,[23] and (c) use a pointing gesture to call attention to some object of interest.[24] What is significant about these findings is *not* that infants can apprehend a pointing gesture as a form of intentional communication—dogs may be capable of that[25]—but rather that infants use pointing to establish a shared perspective with another, to converge on a common object of attention. In other words, the thing *pointed at* is *not* the *goal* of the infant's act of pointing. Rather, the infant's intention is her interlocutor's *attending* to that thing *with her*, i.e., as a shared point of reference. As Henrike Moll and Michael Tomasello explain, "[in infants] pointing is used . . . not just in order to obtain certain objects via helpful adults as social tools, but with the motivation to help/inform others or to just jointly attend to things in the world with them."[26] The pointing gesture is an early form of the child's attempt to forge the shared perspective necessary for persons to delimit a topic of inquiry, as well as an early form of inquiry into the explanation of things or events, including actions.

Thus far, I have discussed one manifestation of a human capacity for joint attention—i.e., the practice of pointing—that enables young children to request and provide explanations of action. Yet, the studies discussed above depended not just on the capacity of a child to raise her index finger, but also on her apprehension of commonalities and differences in the experiential background of her interlocutors. The child points something out to us not only because she wants *us* to attend to the object *with her* but also because the object may be novel or of particular interest *to us*. In the act of pointing, the child shows some awareness of the interpersonal setting in which the communication takes place: "the pointing gesture does not just indicate some spatial location, but instead it already contains a certain perspective from which the indicated object or location is to be viewed."[27] What is interesting here is not so much the practice of pointing, but the child's apprehension of a shared frame of reference, a shared perspective that invests the object pointed at with a particular meaning ("look, here is the toy you liked").

In another set of studies, authors were able to demonstrate that two-year-olds can apprehend differences in the experiential backgrounds of persons they encounter and act accordingly.[28] They have the capacity to act in light of an adult's preferences and the truth or falsity of her beliefs, and to adjust their speech to adults on the basis of prior interaction with the adult in question. For example, eighteen-month-olds who saw that an adult did not witness the removal of an object from a box would not help the adult to open the box when she went back to open it, but rather would retrieve the object from its new location and give it to the adult.[29] Where the children saw that an adult did witness the removal of an object from a box, they would proceed to help her open it, suggesting that the child "believed" there must be something else of interest to the adult in opening the box.[30] Likewise, when an adult excitedly asked for a toy, one-year-olds would select a toy with which the adult had no prior engagement. This not only shows that the child acted on the basis of her apprehension of her interlocutor's prior engagement with the unselected toys, but also her understanding that an expression of excitement, in this experiential frame, *means* something like "can I see the new toy?"

The point here is that even prior to linguistic competency, the child has already begun to develop the capacity to make sense of actions by appeal to a wider experiential frame inhabited by the child and those around her. The *meaningfulness* of certain actions and activities—their purpose or intention—is drawn from the child's continuous interaction with others. The shared point of reference here is not specified by some intentional act (e.g., by pointing to the object in question), but rather by the child's cognizance of the experiential frame in which the action occurs. This is what Moll and Tomasello have identified as the joint-attentional frame,[31] which invests action with a joint meaning and point of reference based on the "intersubjectively shared background of prior experiences."[32] So, within the context of a clean-up frame, the act of pointing to a ring means "put that away," whereas in the context of a play frame, the same act of pointing means "stack it on this ring post."[33] In this regard, the "referent of the pointing gesture thus varies as a function of the joint attentional frame in which the pointing is anchored."[34] The child apprehends how *we* have related, or are presently relating, to some object of attention and acts accordingly.

We could say that the child's cognition of another's intentional action is a function of her apprehending the intentions, purposes, or meanings an action *can have* given the wider experiential frame in which it takes place.[35] The child directly encounters the intentionality of others because she already inhabits a social world invested with meaning and purpose on account of her continuous interactions with purposeful, meaning-driven adults. It is sensible for a child to point to a toy's location for another person against the background of that person's displaying a preference for that toy and not knowing its new location. This shows the child's ability to apprehend the "topic" of her interaction with the adult. The child's cognition takes place in a shared, experiential frame ("we are playing with toys") that constrains and specifies what it makes sense for the adult to do. Given her apprehension of this object of joint engagement, it is not sensible for the child to point to the wall (as opposed to a toy). It would be like someone who understood the game of chess to ask a player, "why are you rearranging those pieces of wood?" Understanding another's actions and intentions depends on a broader apprehension of "what-we-are-doing" and the conventions that delimit and specify that mode of joint engagement.

Consider the first stirrings of language in a young child. In learning to speak, the child must learn a new way that *we* interact with another. This is not something learned *tabula rasa,* as the above studies indicate. At the time in which children typically exhibit language use, they have already learned how to inhabit and navigate between multiple frames of social interaction, developing some understanding of the norms governing different kinds of interaction (e.g., "Mommy and me" v. "Daddy and me") and the meanings of different forms of expression (e.g., excitement v. anger). A child has already become adept at identifying normative patterns of communication with her caregivers—if I smile, Mommy smiles; if I pull Daddy's nose, he makes a funny sound.[36] When these normative patterns are broken, predictable results ensue—the child flies into an inconsolable rage, or, more cunningly, engages in some "flirtation" with Daddy to coax him into the "nose game."[37] These are very localized instances of a child's acting in light of an experiential background with others that is *treated* by the child (and her interlocutors) as normative in some sense. Nevertheless, they illustrate that a pre-linguistic child already inhabits an interactional history with caregivers and non-caregivers that invests future interactions with certain norms and expectations. The child already possesses what we might call a normative sense, an apprehension of the intentions and norms expected to govern these different forms of interaction.

It is no surprise, then, that children as young as two will join an adult in systematically misnaming objects, e.g., calling a dog a cat.[38] The fun of this game is parasitic on the conventional boundaries that it transgresses and so provides evidence of a child's understanding of the normative frame that makes it "fun" *as* a violation of expectation. As Tomasello and Rakoczy put it,

> [T]his kind of play with the conventional use of things—in a way that clearly indicates the child's understanding of the convention and its breaking—illustrates that children participate in the use of symbols with some kind of reflective understanding of their conventional/normative dimension, how one ought to use them under normal circumstances.[39]

So, even at a very young age, children have a capacity for joint action that presupposes knowledge of a norm, of a way "*we* do things." This capacity has been demonstrated even when the norm in question is completely novel. In one study, an adult demonstrated a completely new kind of activity for children, who, in turn, corrected a puppet who did not conform to the norm provided by the adult demonstration.[40] The children were able to see that the puppet was not doing the action in the way *we* do it. This knowledge, of course, presupposed the normativity of their prior interactions with the adult, i.e., the experiential background that frames the point of the interaction between the children, the adult, and the puppet. When, at thirty months, children begin to formulate questions about human activity as well as to provide more expansive explanations of action, they are building on this pre-linguistic capacity to apprehend the normative frame in which human actions take place.[41] These explanations refer to a person's intentions as well as social conventions (e.g., the rules of a game) that govern the action being explained.[42]

In sum, verbal explanations of action and intention, as they occur in young children, are reflective of an already-exercised pre-linguistic capacity to apprehend the normative, experiential, joint-attentional frame that infuses action with a collective purpose and meaning. This form of "we cognition"—how "we" relate to shared points of reference—is what scholars have identified as a capacity for collective intentionality.[43] Through continuous interaction with caregivers and non-caregivers, children enter into increasingly complex "circles of communication" that provide them with this shared, experiential background that provides norms of expectation, expression, and action for future interactions.[44] Although this experiential background is not directly cognized by children, it nonetheless provides the social, normative frame that makes her cognition-specific forms of intentionality and action seem almost effortless. Within the frame of a particular game, for example, the potential *meanings* of some adult's action are significantly constrained. As long as the child apprehends her actions within this frame, the adult's intentions will be completely *transparent*. The frame is normative inasmuch as it provides the explanatory grounds for the rectitude of the child's explanation or description of the adult's action.[45] Yet, prior to the use of speech, it is difficult for us to query the operative frame within which the child apprehends another's intentions. It is precisely through this form of inquiry, however, that moral agreement and disagreement surface. For such an inquiry enables us to challenge the adequacy of *this* normative frame as explanatory grounds for *this* action (e.g., "I understand you were playing a game, but you should have been doing your homework," or "You didn't just eat the cookie—you disobeyed"). While such an inquiry can achieve tremendous levels of complexity that demand a mastery of the language, it nonetheless has its seedbed in the cognitive capacities exhibited in early childhood.

THOMAS AQUINAS AND THE EVOLUTION OF PRACTICAL WISDOM

Before offering some concluding remarks on the wisdom of early childhood, let me briefly summarize the arguments of the previous sections. In the first section, I challenged the theory of mind paradigm that exercises considerable influence over researchers working on cognitive development in children. I explained that the understanding of cognition, and its relationship to human action and intention, held by advocates of this paradigm presupposes a crude dualism between unobservable mental states and observable behaviors that can neither be sustained philosophically nor demonstrated empirically. This is because the system of inferences that constitutes a young child's theory of mind cannot bridge the gap between observable, non-intentional behaviors and unobservable, intentional mental states without admitting that the intentionality of a behavior is, in some minimal sense, known non-inferentially. Indeed, I argued that the knowledge one has of another's intentional action depends upon its being known through a socially mediated frame, i.e., as part of some recognizable practice "we" do. It is only within that frame that our activities can be seen as meaningful and purposive, mistaken or wrong. I identified this mode of cognition as collective intentionality, which I put forward as an alternative paradigm for understanding cognitive development in children.

In the second section, I provided empirical evidence for collective intentionality in young children that demonstrated how, even in infancy, children forge connections with caregivers and non-caregivers that situate their ongoing interactions in a joint-attentional frame. Following other scholars, I noted how this joint-attentional frame provides explanatory grounds for a child's relative ease in identifying human intentions and purposes within the context of joint activity. As these shared, experiential frames expand, a child becomes ever more capable of identifying, introducing, insisting on, and rejecting conventional patterns of human action. In this respect, young children acquire a normative perspective on human action even prior to their ability to express that perspective through language. Although this perspective cannot be directly queried until the child learns to express herself linguistically, I provided some evidence from studies that showed how young children's cognition of the actions and intentions of adults tracks the normative boundaries set by their prior interaction and engagement. Our cognitive capacity is social and normative all the way down, manifesting itself in the earliest stages of cognitive development.[46]

What does all this have to do with the evolution of practical wisdom? With Thomas Aquinas, we can concede that practical wisdom is only partial, restricted to the knowledge of the-good-for-human-beings rather than the good of the whole universe.[47] The practically wise person is one who possesses the virtue of prudence, which is exhibited in her *application* of this knowledge to particular cases in her everyday life. This knowledge is not innate, of course, but acquired through learning and experience.[48] This is why Aquinas claims that one cannot attain practical wisdom without docility, a readiness to be taught by others with greater experience.[49]

Now, we acquire wisdom in the practical domain by coming to grasp the internal relation between the goal of a practice and the standards of evaluation shared by its practitioners.[50] To grasp the inner teleology of a practice is not only to know its basic ground rules, but also to be able to "reason to conclusions in light of them . . . express them accurately in speech, defend them against objections, and determine their very role as principles."[51] Yet, unlike the highest form of wisdom (i.e., metaphysics), in which the relationship between the principles and their conclusions is necessary, the relationship between the rules of a social practice and its goal is contingent. The relationship between the rules of a craft and the excellence at which the craft aims is *adaptable* to changes in circumstance. In years past, a wise warrior would not have required knowledge of the strategic principles that inform practices of "guerilla warfare" and "counter-insurgency." Today, however, one could not be a wise warrior without an understanding of these principles. In other words, what counts as practical wisdom in a particular domain of activity is contingent and variable. It is so not only because of changes in circumstance beyond the control of practitioners, but also due to changes in the conception of the standards that regulate particular social practices. Because the aims of a social practice, and the standards of evaluation framed in light of those aims, are subject to adaptation and change, we cannot expect to acquire any particular form of practical wisdom without undergoing tutelage by those already proficient in the practice.

I have argued that the cognitive capacities of early childhood are the seedbed out of which fully-fledged normative discourse eventually emerges. This is because, by virtue of their capacity for collective intentionality, children already apprehend human activities through the prism of a particular, normative frame. Prior to language, the child is not able to explain her activities by explicit appeal to this frame. As a result, the prelinguistic child cannot be said to possess practical "wisdom" in the sense of being able to explain and justify the internal relation between the goal of practice and its standard of evaluation. Nevertheless, the fluidity and ease with which the young child engages in all sorts of human practices, and identifies all kinds of reasons-for-acting (i.e., intentions), presupposes her capacity to learn, utilize, and eventually master the deployment of the normative frames that make such reasons transparent (and so open to revision and critique). In this, the child is a natural anthropologist, primed to both inhabit and navigate between different kinds of social norms and practices. Therein, I think, lies her inchoate form of practical wisdom.

NOTES

1. For a readable introduction to the concept of collective intentionality and its relation to human cognition, see Michael Tomasello and Hannes Rakoczy, "What Makes Human Cognition Unique? From Individual to Shared to Collective Intentionality," *Mind & Language* 18.2 (2003): 121–47.
2. David Premack and Guy Woodruff, "Does the Chimpanzee Have a Theory of Mind?" *Behavioral and Brain Sciences* 1.4 (1978): 515.
3. George A. Miller, "The Cognitive Revolution: A Historical Perspective," *Trends in Cognitive Sciences* 7.3 (2003): 141–42.
4. Ibid.
5. For the relevant quotes, see George Graham, "Behaviorism," *The Stanford Encyclopedia of Philosophy*, published May 26, 2000, revised March 11, 2015, https://plato.stanford.edu/archives/fall2016/entries/behaviorism/.

6. Alan Costall and Ivan Leudar, "'Theory of Mind': The Madness in the Method," in *Against Theory of Mind,* ed. Ivan Leudar and Alan Costall (New York: Palgrave Macmillan, 2009), 50.

7. Ibid.

8. Ivan Leudar and Alan Costall, "Introduction: Against 'Theory of Mind,'" in *Against Theory of Mind,* ed. Leudar and Costall, 11.

9. Wes Sharrock and Jeff Coulter make this point in their critique of the philosophical presuppositions of "theory of mind." As they explain, if we can only observe physical movements, we will be incapable of directly observing everyday activities like "crossing the street" or "looking in X's direction" since "such descriptions already involve 'intentional' characteristics." On the "theory of mind" account, "we can only observe a body positioned somewhere between one side of the street and the other, with its legs moving. We cannot even assume that the legs are moving so as to transport the rest of the body in any direction: 'crossing the street' anticipates further movements of the body, and characteristically indicates the directedness [read: intentionality] of the mover's steps." Wes Sharrock and Jeff Coulter, "'Theory of Mind': A Critical Commentary Continued," in *Against Theory of Mind,* ed. Leudar and Costall, 67.

10. John O'Callaghan, "How Not to Think About Action, Intention, and Double Effect: The First Person," in *Intention and Double-Effect: Theoretical and Practical Perspectives,* ed. Craig Iffland and John O'Callaghan (Notre Dame, IN: University of Notre Dame Press, forthcoming).

11. Costall and Leudar, "Madness," 51.

12. Carolyn A. Schult and Henry M. Wellman, "Explaining Human Movements and Actions: Children's Understanding of the Limits of Psychological Explanation," *Cognition* 62.3 (1997): 318.

13. Ibid.

14. The expression "forms of life" is taken from Ludwig Wittgenstein, *Philosophical Investigations,* trans. G. E. M. Anscombe, P. M. S. Hacker and Joachim Schulte, ed. P. M. S. Hacker and Joachim Schulte, 4th ed. (Oxford: Wiley-Blackwell, 2009), § 23, 241.

15. As Ivan Leudar and Alan Costall so eloquently put it, "[o]ur everyday use of language is a practical matter. As in natural science, we can use everyday language to describe, explain and predict actions of others. These three are, however, just some amongst many speech acts, none of which are actions of solitary observers oriented to but disengaged from the world of others, but rather conversational activities done with other people." Ivan Leudar and Alan Costall, "On the Historical Antecedents of the Theory of Mind Paradigm," in *Against Theory of Mind,* ed. Leudar and Costall, 23.

16. Mattia Gallotti and Chris D. Frith, "Social Cognition in the We-Mode," *Trends in Cognitive Sciences* 17.4 (2013): 162.

17. On the anthropological dimensions of Wittgenstein's understanding of "forms of life," see Gordon P. Baker and Peter Hacker, *Wittgenstein: Rules, Grammar and Necessity: Essays and Exegesis of §§ 185-242* (Oxford: Wiley-Blackwell, 2009), 218–23.

18. I take the notion of "frame" from Rom Harré, "Norms in Life: Problems in the Representation of Rules," in *Jerome Bruner: Language, Culture and Self,* ed. David Bakhurst and Stuart Shanker (London: Sage Publications, 2001), 155–56.

19. Henrike Moll and Michael Tomasello, "Cooperation and Human Cognition: The Vygotskian Intelligence Hypothesis," *Philosophical Transactions of the Royal Society of London B: Biological Sciences* 362.1480 (2007): 639.

20. Michael Tomasello and Malinda Carpenter, "Shared Intentionality," *Developmental Science* 10.1 (2007): 121.

21. Henrike Moll, "Human Specific Forms of Cognition Prior to Judgments," *Grazer Philosophische Studien* 86 (2012): 238.

22. David Premack and Ann James Premack, "Infants Attribute Value to the Goal-Directed Actions of Self-Propelled Objects," *Journal of Cognitive Neuroscience* 9.6 (1997): 849.

23. Ulf Liszkowski et al., "Pointing Out New News, Old News, and Absent Referents at 12 Months of Age," *Developmental Science* 10.2 (2007): F1–F7.

24. Ulf Liszkowski et al., "Twelve-Month-Olds Point to Share Attention and Interest," *Developmental Science* 7.3 (2004): 297–307.

25. Although dogs *can* apprehend a pointing gesture as indicative of the presence of food, studies have shown that dogs will not follow a pointing gesture if they know it does not direct them to the hidden treat or where they have not previously found food in the same room. See Linda Scheider et al., "Do Domestic Dogs Interpret Pointing as a Command?" *Animal Cognition* 16.3 (2013): 361–72.

26. Moll and Tomasello, "Cooperation," 643.

27. Ibid., 644.

28. The authors conclude that a child's attentiveness to another's experiential background enables her to achieve shared points of reference with others, which is essential for increasingly complex forms of communication and cooperation. See Henrike Moll and Derya Kadipasaoglu, "The Primacy of Social over Visual Perspective-Taking," *Frontiers in Human Neuroscience* 7 (2013): 193.

29. Ibid., 189.

30. Ibid.

31. Moll and Tomasello, "Cooperation," 644.

32. Moll and Kadipasaoglu, "The Primacy of Social over Visual Perspective-Taking," 190.

33. Kristin Liebal et al., "Infants Use Shared Experience to Interpret Pointing Gestures," *Developmental Science* 12:2 (2009): 264–71.

34. Moll and Tomasello, "Cooperation," 644.

35. By "goal," I mean anything that explains "why" there is an "action." To specify the aim of an action is to offer a kind of explanation of the action.

36. Stanley I. Greenspan and Stuart Shanker, *The First Idea: How Symbols, Language, and Intelligence Evolved from Our Early Primate Ancestors to Modern Humans* (Cambridge: Da Capo Press, 2006), 59.

37. Greenspan and Shanker describe how the child's apprehension of different normative patterns of interaction allows her to negotiate interactions with caregivers by deploying conventional patterns of recognition in anticipation of a particular pattern of emotional response. Ibid., 33.

38. Tomasello and Rakoczy, "What Makes Human Cognition Unique?" 128.

39. Ibid., 128–29.

40. Tomasello and Carpenter, "Shared Intentionality," 124.

41. Anne K. Hickling and Henry M. Wellman, "The Emergence of Children's Causal Explanations and Theories: Evidence from Everyday Conversation," *Developmental Psychology* 37.5 (2001): 668–83.

42. Hickling and Wellman, "Emergence," 671.

43. Galloti and Frith, "We-Mode," 160.

44. I take the term "circle of communication" from Greenspan and Shanker, *The First Idea*, 42.

45. Here again I rely on Harré, "Norms in Life," 155–56.

46. On this point, I am indebted to Andrea Kern and Henrike Moll, "On the Transformative Character of Collective Intentionality and the Uniqueness of the Human," *Philosophical Psychology* 30.3 (2017): 315–33.

47. Thomas Aquinas, *Summa Theologica*, trans. Fathers of the English Dominican Province, 3 vols. (New York: Benziger Bros., 1947–1948), 2a2ae 47.2 *ad* 1.

48. Ibid., 47.15.

49. Ibid., 49.3.

50. For this understanding of practice, I am indebted to Alasdair MacIntyre, *After Virtue: A Study in Moral Theory*, 3rd ed. (Notre Dame, IN: University of Notre Dame Press, 2007), 181–204. On the internal relation of a rule to the practice it regulates, see Baker and Hacker, *Wittgenstein*, 46–55.

51. Stephen Brock, *The Philosophy of Saint Thomas Aquinas: A Sketch* (Eugene: Cascade, 2015), 95.

BIBLIOGRAPHY

Aquinas, Thomas. *Summa Theologica*. Translated by Fathers of the English Dominican Province. 3 vols. New York: Benziger Bros., 1947–1948.

Baker, Gordon P., and Peter Hacker. *Wittgenstein: Rules, Grammar and Necessity: Essays and Exegesis of §§ 185-242*. Oxford: Wiley-Blackwell, 2009.

Brock, Stephen. *The Philosophy of Saint Thomas Aquinas: A Sketch*. Eugene: Cascade, 2015.

Costall, Alan, and Ivan Leudar. "'Theory of Mind': the Madness in the Method." In *Against Theory of Mind*, edited by Ivan Leudar and Alan Costall, 39–55. New York: Palgrave Macmillan, 2009.

Gallotti, Mattia, and Chris D. Frith. "Social Cognition in the We-Mode." *Trends in Cognitive Sciences* 17.4 (2013): 160–65.

Graham, George. "Behaviorism." *The Stanford Encyclopedia of Philosophy*. Published May 26, 2000. Revised March 11, 2015. https://plato.stanford.edu/archives/fall2016/entries/behaviorism/.

Greenspan, Stanley I., and Stuart Shanker. *The First Idea: How Symbols, Language, and Intelligence Evolved from Our Early Primate Ancestors to Modern Humans*. Cambridge: Da Capo Press, 2006.

Harré, Rom. "Norms in Life: Problems in the Representation of Rules." In *Jerome Bruner: Language, Culture and Self*, edited by David Bakhurst and Stuart Shanker, 150–66. London: Sage Publications, 2001.

Hickling, Anne K., and Henry M. Wellman. "The Emergence of Children's Causal Explanations and Theories: Evidence from Everyday Conversation." *Developmental Psychology* 37.5 (2001): 668–83.

Kern, Andrea, and Henrike Moll. "On the Transformative Character of Collective Intentionality and the Uniqueness of the Human." *Philosophical Psychology* 30.3 (2017): 315–33.

Leudar, Ivan, and Alan Costall. "Introduction: Against 'Theory of Mind.'" In *Against Theory of Mind*, edited by Ivan Leudar and Alan Costall, 1–18. New York: Palgrave Macmillan, 2009.

———. "On the Historical Antecedents of the Theory of Mind Paradigm." In *Against Theory of Mind*, edited by Ivan Leudar and Alan Costall, 19–38. New York: Palgrave Macmillan, 2009.

Liebal, Kristin, Tanya Behne, Malinda Carpenter, and Michael Tomasello. "Infants Use Shared Experience to Interpret Pointing Gestures." *Developmental Science* 12.2 (2009): 264–71.

Liszkowski, Ulf, Malinda Carpenter, Anne Henning, Tricia Striano, and Michael Tomasello. "Twelve-Month-Olds Point to Share Attention and Interest." *Developmental Science* 7.3 (2004): 297–307.

Liszkowski, Ulf, Malinda Carpenter, and Michael Tomasello. "Pointing Out New News, Old News, and Absent Referents at 12 Months of Age." *Developmental Science* 10.2 (2007): F1–F7.

MacIntyre, Alasdair. *After Virtue: A Study in Moral Theory*, 3rd edition. Notre Dame, IN: University of Notre Dame Press, 2007.

Miller, George A. "The Cognitive Revolution: A Historical Perspective." *Trends in Cognitive Sciences* 7.3 (2003): 141–44.

Moll, Henrike. "Human Specific Forms of Cognition Prior to Judgments." *Grazer Philosophische Studien* 86 (2012): 235–45.

Moll, Henrike, and Derya Kadipasaoglu. "The Primacy of Social over Visual Perspective-Taking." *Frontiers in Human Neuroscience* 7 (2013): 188–96.

Moll, Henrike, and Michael Tomasello. "Cooperation and Human Cognition: The Vygotskian Intelligence Hypothesis." *Philosophical Transactions of the Royal Society of London* 362.1480 (2007): 639–48.

O'Callaghan, John. "How Not to Think About Action, Intention, and Double Effect: The First Person." In *Intention and Double-Effect: Theoretical and Practical Perspectives*, edited by Craig Iffland and John O'Callaghan. Notre Dame, IN: University of Notre Dame Press, forthcoming.

Premack, David, and Ann James Premack. "Infants Attribute Value to the Goal-Directed Actions of Self-Propelled Objects." *Journal of Cognitive Neuroscience* 9.6 (1997): 848–56.

Premack, David, and Guy Woodruff. "Does the Chimpanzee Have a Theory of Mind?" *Behavioral and Brain Sciences* 1.4 (1978): 515–26.

Scheider, Linda, Juliane Kaminski, Josep Call, and Michael Tomasello. "Do Domestic Dogs Interpret Pointing as a Command?" *Animal Cognition* 16.3 (2013): 361–72.

Schult, Carolyn A., and Henry M. Wellman. "Explaining Human Movements and Actions: Children's Understanding of the Limits of Psychological Explanation." *Cognition* 62.3 (1997): 291–324.

Sharrock, Wes, and Jeff Coulter. "'Theory of Mind': A Critical Commentary Continued." In *Against Theory of Mind*, edited by Ivan Leudar and Alan Costall, 56–88. New York: Palgrave Macmillan, 2009.

Strauss, Sidney, Margalit Ziv, and Adi Stein. "Teaching as a Natural Cognition and its Relations to Preschoolers' Developing Theory of Mind." *Cognitive Development* 17.3 (2002): 1473–87.

Tomasello, Michael, and Malinda Carpenter. "Shared Intentionality." *Developmental Science* 10.1 (2007): 121–25.

Tomasello, Michael, and Hannes Rakoczy. "What Makes Human Cognition Unique? From Individual to Shared to Collective Intentionality." *Mind & Language* 18.2 (2003): 121–47.

Wellman, Henry M., and Kristin H. Lagattuta. "Theory of Mind for Learning and Teaching: The Nature and Role of Explanation." *Cognitive Development* 19.4 (2004): 479–97.

Wittgenstein, Ludwig. *Philosophical Investigations*. Translated by G. E. M. Anscombe, P. M. S. Hacker and Joachim Schulte, edited by P. M. S. Hacker and Joachim Schulte. 4th edition. Oxford: Wiley-Blackwell, 2009.

IV

Evolving Wisdom as Virtue

EIGHT

Change and Constancy in the Nature of Wisdom over Time

Adam Willows

In this chapter I discuss the way in which existing theological ideas about both wisdom and human nature can be applied to our understanding of our evolutionary past. In order to provide a unified example of how such an understanding might develop, I focus on the work of Thomas Aquinas as a representative of the theological tradition and its commitments. I begin by examining what the ascription of wisdom to our ancestors would mean for our understanding of their nature and our relationship to them. I show that Aquinas thinks wisdom requires the presence of an intellectual soul, meaning that any wise ancestors would share our human nature. I then discuss the problems that arise from the suggestion that our species' evolutionary development included the transition from one kind of soul to another. I argue that it does not entail the drawing of negative distinctions, either between humans and other creatures or between different humans. I also discuss how understanding the evolution of wisdom requires an explanation of changes in the soul as well as accounts of the evolution of the body. I show that Aquinas holds the transition between types of soul to be a very different process to that of our species' gradual material change, but argue that the two are not incompatible. Having discussed the difference between wise and non-wise creatures and the nature of the movement from one to the other, I conclude by trying to understand possible differences between us and our ancestors by looking at Aquinas' account of the different levels and types of wisdom. I show that it is reasonable on Aquinas' view to expect differences to be present, but not for these differences to necessarily be to modern humans' advantage.

There are many ways that wisdom appears in Aquinas' thought. In humans, the virtue of wisdom is the consideration of the "highest causes"—the ability to grasp the ultimate meaning and explanation of things.[1] Most importantly, this kind of wisdom necessarily directs us to God, who is the ground of all things and for whom wisdom is therefore self-knowledge, possessed in a different and more complete way.[2] God creates according to God's wisdom, meaning that that is a further sense in which the divine wisdom is expressed through, and present in, all creation.[3] Throughout this chapter I concentrate on the first kind of wisdom—the virtue and its expression in humans—and this is the meaning that should be understood unless otherwise stated.

I have said what the purpose of this chapter is, but, before beginning, I also want to note what it is not. This chapter is not an attempt to argue for or against the claim that wisdom was present in our ancestors. Rather, it seeks to understand the theological implications of making such a claim by tracing the connections between wisdom and theological thought on other aspects of human nature and creation. Nor is this chapter an attempt to advocate for Aquinas' understanding of wisdom and humanity, or to make any empirical claims on that basis.[4] I am on the whole sympathetic to Aquinas' position on these matters, and much of his argument concerning them appears here. Nevertheless, my goal is simply to follow the train of his thought to draw out a representative account of how theological theory might engage with the evolution of wisdom.

THE POSSESSION OF WISDOM

I hope to give a theological perspective on this question: How might wisdom be different for other species such as our hominin ancestors? I want to understand what parts or outworkings of wisdom (if any) we might expect to change across species, and what we might expect to stay the same. In asking this, I face an immediate problem: how can I expect to identify cross-species changes in wisdom when we only have a single point of reference? I cannot compare wisdom across species without species to compare across; and we only have one currently available species thought to exhibit wisdom—us.

In resolving this difficulty I intend to look not back, to our distant past, but up. Theologians do in fact identify another species that possesses wisdom. Besides ourselves, who are only partly and inconsistently wise, and God, who is wisdom itself, there are other rational beings above us in the theological hierarchy of being.[5] These are the angels. I suggest that by comparing wisdom in the angels to wisdom in humans, we can produce a theological understanding of how wisdom might change across species. Aquinas is particularly useful here, since as well as writing at length on the virtue of wisdom he earned the name "the Angelic Doctor" for his work on angels.

In this first section I want to look at features and consequences of wisdom that remain constant over time. If both we and our ancestors were wise, what commonalities would Aquinas identify between us? To begin, it is worth asking why angels and humans can be wise but no other creatures can. For Aquinas, the essential distinction is that both possess an intellectual, or rational, soul. Plants and animals possess lower souls which depend on the body and are therefore mortal.[6] Intellectual souls, on the other hand, are not corporeal.[7] Angels are truly intellectual creatures, whereas humans—and only humans—are composed of a body and an intellectual soul.[8] A virtue is a perfected power of the soul, so in order to possess a virtue you must possess the kind of soul to which it pertains.[9] Wisdom is an intellectual virtue, so in order to be wise a creature must have an intellectual soul.[10] Only humans and angels meet this criterion. Wisdom, then, cannot be exclusively related to the material, depending as it does on an immaterial soul.

Here I think there are two interesting points to be made regarding our ancestors. Firstly, humans are the *only* kind of creature that unites both corporeal matter and intellectual soul.[11] That is simply what a human is: matter plus soul.[12] So if our ancestors were wise then they must, on Aquinas' terms, have been human. Interestingly, the reason for this is that he thinks that if intellect is to be united to matter, it can only be united to the kind of matter that will allow it to be properly expressed. This is one of the key differences between us and angels.[13] Angels, as pure intellect, simply receive understanding. Humans are united to matter, meaning that we are limited to achieving our perfection through the medium of matter. In other words, to gain understanding, humans need information acquired through sense data. If intellect is going to be united to matter, it needs to be in a form that allows it to effectively receive and interpret sense data. That, Aquinas thinks, is what the human body is: the physical form best suited to gaining understanding from the world.[14] He places a lot of importance on the sense of touch; we might instead identify higher encephalization and opposable thumbs—but the principle is the same.[15]

It is worth distinguishing here between the intellectual soul and the physical traits that allow complex cognition. In humans, they are found together; and certainly, a body sufficiently suited to performing this cognition will be suited to an intellectual soul. But this does not rule out the possibility that other species, and particularly our hominin ancestors, could have developed these physical traits at least to some extent—social behavior and tool use among other species today are good examples of this.

However, this distinction cannot be pushed too far. Aquinas' understanding of the relationship between the body and soul is essentially hylomorphic. Both are essential for the existence and nature of a human; but unlike a dualist conception of the relationship between mind and body, there is no easy separation to be made between the two. To say humans are made of body and soul is not, for Aquinas, to say that we are a mere combination of the two: "Instead, the hylomorphic tradition construes a human being as a single substance, a thinking, living creature resulting from a soul configuring matter."[16] The expression of wisdom in humans depends both on an intellectual soul and on a suitable body of which that soul is the form; it is out of this unity of the material and immaterial that a human being is produced. The understanding of wisdom in human history, then, will for the Thomist be one which takes account of the evolutionary adaptations which allowed for interpretation of sense data, and the divine infusion of the intellectual soul. Both together result in human beings with the capacity for wisdom.

What does this mean for our understanding of our ancestors? Firstly, theologically speaking, there is no room for halfway houses. If they had the potential for wisdom then they were as human as you or I. This does not mean denying physiological or behavioral differences between us and our ancestors; it means denying that human nature, theologically understood, is constituted by those differences. Secondly, though, there will be a cut-off point. Not every physical form is sufficiently suited to receiving and interpreting sense data to be joined with an intellectual soul. This is where humanity—and wisdom—stops. I do not care to speculate where this boundary might be; perhaps (even probably) it would be a messy line with plenty of fringe cases. Nevertheless, it is quite clear that Aquinas would think that there is one. He considers the possibility of another animal possessing an intellectual soul, and says that any such animal's form would have to be so closely related to the human in order for this to be possible that there would be no real difference between them: "there would be no specific difference between the animal so constituted and man."[17] Any ancestor, then, must be very like us to have the potential for wisdom; but those that do have that potential will be just as much embodied, intellectual creatures—just as human—as us.

The next point I want to make is that, per the discussion above, anything with the potential to be wise necessarily has an intellectual soul. Intellectual souls are necessarily immortal (in Aquinas' terms, incorruptible).[18] So if we talk of our ancestors as wise we must include them in our eschatology. This also, for Aquinas, follows from their being human; but this drives the point home. Aquinas' eschatology is fundamentally a "wisdom eschatology."[19] For potentially wise beings to be left out is, to him, a contradiction. He directly says that it would not make sense for there to be a rational soul unable to participate in beatitude.[20] Christ's Passion "was superabundant and sufficient satisfaction . . . for the whole human race."[21] Aquinas thinks that faith in Christ was possible before the Incarnation and that Christ's sacrifice was efficacious for those people.[22] Finally, if our ancestors had the capacity for wisdom then they will partake in the bodily resurrection.[23] Put simply, if our ancestors had the capacity for wisdom, then Christ died for them.

BOUNDARIES AND DIVISIONS

Aquinas thinks the ascription of wisdom to an embodied creature means the ascription of an intellectual soul, humanity, and all that that entails. There is a boundary to this, though: not all physical forms can be joined with an intellectual soul. As I say above, I do not think it worthwhile to try to identify exactly where the cut-off for possession of an intellectual soul, and therefore wisdom, might be. But even the suggestion that this boundary exists may be enough to cause uneasiness. It appears that this position could be divisive in two ways: first, by drawing a line between those with an intellectual soul and those with a sensitive or vegetative soul, it creates too much of a sharp distinction between humans and non-humans. This runs the risk both of diminishing the importance of non-human creatures, and of excessively elevating humans. It is this kind of approach that can encourage theologies of dominion rather than stewardship, which, unchecked, may be taken to justify mistreatment of the natural world.[24]

Second, as well as dividing humans from other creatures, this position might divide human from human. One of the marks of an intellectual soul is the capacity for rationality; but there are many in whom the ability to reason is reduced or not present at all. If humans result from the unity of a body and an intellectual soul, does this position suggest that those who do not exhibit rationality are in some way less human? If this is the case, then the idea of a boundary ought to be rejected on moral grounds. It would devalue those who exhibit reduced or no rationality, to many of whom we owe a special duty of care.

Although Aquinas thinks a boundary exists, I think he would deny that the negative implications discussed above apply. With regard to the division between rational and non-rational creatures, it is evident that he believes there to be a difference in goodness between the two. This is because different creatures reside at different levels on the hierarchy of being, the order in which created things participate in and reflect the goodness of God.[25] Humans are the lowest of the creatures with intellectual souls, but they possess more goodness than non-rational creatures. Given this, how is Aquinas to resist the conclusion that his position diminishes the importance of other creatures? The first clue is that Aquinas thinks that these differences result in *more* overall goodness, not less. "Therefore, as the divine wisdom is the cause of the distinction of things for the sake of the perfection of the universe, so it is the cause of inequality. For the universe would not be perfect if only one grade of goodness were found in things."[26] Here inequality is not a negative thing; the universe is like a work of art, with differences coming together to contribute to the perfection of the whole. Thus Aquinas thinks it would be wrong to think of the lower orders of creation as *worse* in some way. In fact, he says there is no prior inequality of merit between different creatures.[27] By this he appears to mean that each creature is not somehow composed of less valuable material. He does believe there to be inequality between different species, but this only appears in the comparison between them; one could not identify any particular species by itself as defective. Each creature is precisely what it needs to be to represent the goodness of God in its particular way.

Nor should this distinction result in humans mistreating or dismissing the rest of creation. For Aquinas, this would be entirely wrongheaded; in fact, the opposite is true. It is this difference that allows humans to see *better* the value of creation and to love it more. The division between rational and non-rational creatures brings an increase in awareness of the value of creation, its particular goods, and love for the non-rational. Consideration of creatures is an important way in which humans are able to attain their own good and to develop wisdom themselves (although creatures are valuable in themselves, not merely because of their benefit to us).[28] Some of our ancestors far enough back in our evolutionary history would, on Aquinas' terms, lack a physical form suited to an intellectual soul and thus be incapable of wisdom. This is indeed a division, but not one that would imply that those ancestors were not valuable or important. Jame Schaefer notes that "Aquinas depicted each entity as perfect in some innate way . . . To criticize a creature's nature or its natural inclination is, he insisted, an insult to the Creator of nature."[29]

What about the potential for division between humans? In his discussion on the intellectual soul Aquinas explicitly rejects the possibility that a difference in the rational powers between people would result in a difference in nature. He considers the suggestion that "if one man understands better than another, it would seem that they do not belong to the same species."[30] His response is that this kind of difference between people is "due only to the different disposition of matter," and is thus not the kind of difference to cause a difference in nature.[31] So human nature is equally present and equally valuable in all humans. This is, as far as it goes, an answer to the problem. But this response might seem like shifting the goal posts. Very well (we might say), differences in rational capacities cannot cause differences in human nature; but they can themselves be caused by other differences that are due to the "disposition of matter," and *those* differences can be the basis for devaluing a person.

This complaint has a long way to go to be viable. Aquinas is simply making the very obvious point that the makeup of the material world results in differences between people. This, he thinks, is required for individuality.[32] Without it there would be no differences in form and so no distinction between persons. He does think, though, that some people are closer to perfection than others—but not in their nature, as discussed. Instead, it is because they may possess more or less of certain goods, due to their character and sometimes their circumstances. However, there are various ways to lack perfection; all humans do to some extent and not necessarily through any fault of their own. In this context, Aquinas does not seem especially concerned about those who cannot exercise their rational capacities. To understand why, it is important to understand that for Aquinas the *possession* of wisdom is not necessarily attached to the *expression* of wisdom. It is possible, Aquinas thinks, to be wise but for some reason to be unable to act wisely. This is true both of humanity's natural capacity for wisdom and of the divine wisdom granted by grace. Particular bodily circumstances may prevent someone from acting wisely, but these are not essential to the person. In fact, John Berkman argues that because they cannot sin, Aquinas holds children and those affected by illness or injury to be more secure in their wisdom and "closer to heavenly perfection than those who can and do act."[33]

So Aquinas does think that some humans are different because they are rationally impaired. This does not, however, indicate any difference in essential nature and cannot be the basis for drawing any but circumstantial distinctions between humans. Furthermore, even if this kind of distinction is drawn, it appears to suggest that those with restricted rational capacities are *more* perfect, not less. "Though it is natural for moderns to think that questions of mental or physical impairment are of paramount importance, for Aquinas they are not nearly as important as moral or spiritual impairment."[34]

It cannot be denied that the idea that there is essential inequality between creatures and persons runs through Aquinas' thought. As such, it is probably fair to say that the Thomist should be especially alert to any opening of the door to thought that allows the mistreatment of animals or diminishment of persons. What can be denied is that this door is necessarily open. For Aquinas, inequality is not always pejorative, and in the case of humans does not align neatly with particular disabilities. He does indeed think that some people possess less good than others. It is moral good that Aquinas is most concerned about, however, and the primary cause of this difference is not impaired rationality but sin.[35] It is chiefly those who are *morally* worse who Aquinas believes lack important goods. Humans *qua* human are equally valuable; but those who damage themselves by developing a vicious character will necessarily find themselves in a different place to those who cultivate virtue.

I have argued that Aquinas' boundary between those creatures with an intellectual soul and those without does not lead to negative judgments, either of those who lack an intellectual soul or of those who cannot express its capacities. When considering our ancestors this first means that those who lacked an intellectual soul are not thereby devalued. More interestingly, though, he also thinks that intellect, while a good in itself, is not by itself a guide to the wisdom or overall perfection (or imperfection) of an individual. The mere fact that we have greater cognitive capacity than some of our ancestors does not, therefore, mean that we are better than them. In fact, it is quite possible according to Aquinas' position that they were closer to human perfection than we are. In the next section, I want to continue discussing the implications of the boundary Aquinas draws between creatures with and without an intellectual soul. I am going to address the possibility that this boundary is not consistent with what we know about human evolutionary development.

EVOLUTION AND THE APPEARANCE OF THE SOUL

I will return to Aquinas' thought on different modes of perfection later in the chapter. I have argued that Aquinas' view that the intellectual soul is necessary for wisdom need not lead to negative views of non-intellectual creatures or cause division between humans. In this section I want to examine how Aquinas' understanding of the soul and the appearance of wisdom coheres—or does not cohere—with evidence about the evolutionary development of our species.

I have shown that Aquinas thinks it is not possible to say that a particular creature is not human and lacks an intellectual soul simply because they do not possess the capacity to act wisely. The capacity to possess wisdom is essential to humanity because it comes with our intellectual souls. Expressing or acting on that wisdom is not. How then are we to determine a creature's humanity apart from observing wise or rational action? For Aquinas, the answer is simple: it was born of human parents. This does not mean that physical reproduction is what creates humans, however. To say that the human soul was produced by reproduction would be to say that it was material and hence perishable. As an immaterial substance the intellectual soul can only be produced by God alone.[36] Rather, he thinks that physical reproduction creates the conditions for humanity, by arranging matter in a way that allows it to receive the human form: "Man begets his like, forasmuch as by his seminal power the matter is disposed for the reception of a certain species of form."[37]

This might have seemed simple enough to Aquinas, but it appears to introduce a significant element of confusion when attempting to locate human nature in our evolutionary past. Go back far enough and our ancestors are not human, but any attempt to identify the specific point of change will be fruitless. In physical terms, there was no sudden shift from "non-human" to "human." But if we cannot identify whether or not our ancestors were wise simply from their behavior, and we cannot determine a point at which human parentage began, then Aquinas' method of identifying humans—however well it works today—seems completely unsuited to setting a clear boundary between wise and non-wise creatures. Does this not suggest that Aquinas is mistaken, that the capacity for wisdom developed gradually, and that it would be better to do away with the idea of a strict boundary altogether? Note, though, that Aquinas does not think that human parentage by itself is what makes us human. What it does do is provide us with a reliable way to identify humans. I think what this underlines—at least for Aquinas—is the way that human nature does not depend only on the material. Aquinas can accept that even with total physical information about every single ancestor it would not be possible to identify a "start point" for humanity, but still maintain his view that there is such a start point. This is because the location of this start point is not a physical matter at all; it is determined by an immaterial (non-physical) event that is down to God.

This may well seem very strange to a modern reader. Our material forms and corresponding capacities changed so gradually and incrementally. Should the immaterial part of our nature not follow? To Aquinas, though, this position has it backward: "the body depends on the soul, and not the soul on the body."[38] A more accurate perspective would be that all humans, as intellectual creatures, are fundamentally distinct from non-intellectual creatures, but our intellectual souls are joined to bodies in sufficiently diverse ways that there may be material similarity with non-intellectual creatures. Aquinas was of course not aware of creatures as closely related to modern humans as our hominin ancestors, but it is probable that he would not see a gradual transition as a problem for his view. He says that the highest creatures in the lower genera always "come close" to the lowest creatures in the genera above them, and celebrates this as an example of "the marvellous connection of things."[39]

There is, in fact, clear evidence that this would have been Aquinas' position. Although unaware of evolutionary processes, he considers a situation in which gradual physical changes are accompanied by a sudden transition from one kind of soul to another. In the *Summa Contra Gentiles*, he discusses the process of conception and fetal development. The development in question is that of an individual, not a species—but the question he deals with has application to both: how do changes in the nature of the soul correspond to a very gradual material change from a form that cannot receive an intellectual soul to one that can? In evolutionary terms, what would Aquinas say about the souls of our hominin ancestors as they developed into modern humans?

Perhaps the most obvious instinct for a reader familiar with evolutionary processes is to think that the soul gradually developed along with the body, slowly becoming more intellectual and so incrementally developing the capacity for wisdom. However, this appears to be at odds with

Aquinas' view that there is an essential category difference between intellectual souls, which are immaterial, and all other kinds of soul, which are material.[40] He goes on to confirm this by explicitly denying the possibility that one kind of soul could transition to another:

> Now, this theory would involve the consequence that numerically one and the same power is at one time a purely vegetative soul, and afterward a sensitive soul, the substantial form itself thus being perfected successively more and more . . . Now, all these consequences are impossible in nature.[41]

Later he returns to this subject at length in a separate text on the soul.[42] He determines that human development is a situation where lower kinds of soul (vegetal and sentient) are eventually replaced with an intellectual one. The reason this process cannot involve the lower souls slowly developing into the higher is that for one to transition into another would mean that the intellectual soul depended on the material world for its origins. In a study of Aquinas on human life, Fabrizio Amerini discusses what this implies: "if we did admit that the rational soul was an 'evolution' from the sensitive soul (even if by means of an external agent), then either the rational soul would be corporeal and so mortal . . . or else vice versa, the vegetative and sensitive souls would be immortal."[43]

Instead, Aquinas concludes that at the appropriate juncture (he does not say exactly when), the vegetal soul is simply discarded and replaced with a sentient one, which is later replaced in the same way by an intellectual soul: "when the human soul arrives, the substantial form which inhered before is removed."[44] This is not precisely the same situation as evolutionary development of a species, but it does establish that Aquinas saw nothing peculiar about incremental physical change being accompanied by sudden changes in the kind of soul present. When this happens will be down to God, whose direct action is the only thing that can bring an intellectual soul into being.[45] *That* it happens, though, seems to be accepted by Aquinas as the most natural explanation for the way souls change with changing bodies.

Does this mean that Aquinas' theory of soul is simply a product of its time, a best guess unsuited for use in a modern setting aware of evolutionary processes? I do not think this need be the case. The vital point to remember is that Aquinas thinks humans are a unique product of two entirely different kinds of created substance: the material, which forms other creatures and is subject to gradual change and eventual decay; and the immaterial, which is intellectual, subsistent and immutable.[46] We are therefore peculiar creatures, a subject most evident in theological discussion of the Incarnation, where the apparent paradox of the unity of material and immaterial is joined with the further paradox of unity between Creator and creature.[47] Of the Incarnation, Aquinas says "to explain this union perfectly is beyond man's strength . . . in all created things nothing is found so like this union as the union of soul to body."[48] Given this, it should not be too great a surprise that these different but united aspects of a human being operate quite differently from one another.

Amerini suggests that this can be understood with reference to Aquinas' theory of causation. The soul is the formal cause of a creature; it is the thing that determines what that creature is, its particular essence—"dog-ness" or "human-ness." The body is the material cause of a creature—the thing from which a dog or human is made. Modern causal theory is typically limited to only one kind of cause, and from this position Aquinas' explanation of how body and soul unite to make a human being can seem especially strange. Within a medieval causal theory it is less so. Soul and body both cause a human being to exist; but they do this in fundamentally different ways, and there is no reason to expect that the process of causation in one case will be the same in the other. The transition between souls, then, is discontinuous rather than incremental, while on the material plane "there is no observable evidence of any substantial interruption in the process."[49] Just as Aquinas does not think gradual change in the body is replicated in the soul, he need not think that movement from one kind of soul to another would necessarily interfere with that process of gradual change. Aquinas would not think that his theory of soul conflicted with evolutionary processes—simply that it is not governed by them.

I think it is now possible to draw some conclusions about how the idea of a sharp line between intellectual and non-intellectual creatures should be handled by Thomists looking at our ancestors. When discussing human nature in the abstract, the distinction Aquinas draws between those with intellectual souls and those without might prove helpful. It serves to clarify exactly which traits Aquinas thinks constitutes human distinctiveness. This, in turn, will make it easier to identify common ground between us and our ancestors based on our shared humanity. However, when discussing humans and our ancestors in more concrete terms—identifying and studying particular groups, places and times—attempting to replicate the clear line that applies in the abstract will in fact obscure the topic at hand. Given the method Aquinas offers for identifying humans, it would be entirely fruitless to try to precisely locate a human "beginning"; such a beginning is essentially unknowable to us. Instead, it might be better to think of an incremental and perhaps sporadic growth toward wisdom. In at least one sense Aquinas would accept this. His position need not deny that the material changes that eventually suited our ancestors' forms to being joined with an intellectual soul were very gradual. Further, once this union of body and soul did occur, subsequent changes may have allowed improved exercise of the capacity for wisdom, if not changes in that capacity itself. He can also accept that significant variation will only become apparent over extended periods of time.

There is no getting around the fact that Aquinas' position implies a spiritual "leap" from creatures without intellectual souls to those with them; souls do not blend between categories, and are not capable of "intensification and remission."[50] Importantly, though, this spiritual leap need not be accompanied by a corresponding physical one. Material changes can remain as gradual as ever.

LEVELS OF WISDOM

I have talked so far about what it means to be a wise creature in common with our ancestors, the differences between creatures with and without the capacity for wisdom, and the transition in our history from one to the other. Now I want to move on and consider differences between various wise creatures. I will look at levels of wisdom and possible differences in wisdom between us and our ancestors. Wisdom, Aquinas thinks, is an understanding of "first principles" or the true nature of things.[51] Thus a wise doctor truly understands the practice of medicine and what it is to be a doctor, and a fully wise person understands more broadly the purpose and nature of all things. Although humans and angels share a capacity for wisdom, our levels of understanding differ. Angelic knowledge is of the same kind as human knowledge, but is "as the perfect to the imperfect."[52] In fact, there is a ranking of knowledge and wisdom among the angels, as well. Those who are higher in the hierarchy, "beholding the Divine wisdom more clearly, learn more and deeper mysteries in the vision of God, which mysteries they communicate to the lower angels by enlightening them."[53] There is a scale or spectrum of wisdom. Angels have greater wisdom than humans but they also vary among themselves.[54] Humans continue the scale, although lower down. In fact, Aquinas thinks that just as with lower souls, the continuum is pretty seamless; the wisest of us "touch the base of the angelic nature."[55] This spectrum continues among humans themselves; Aquinas thinks that all are loved by God, but between different individuals and at different stages of our lives, we find differing levels of virtue.[56] Wisdom, being one such virtue, varies between us.

I think that this discussion of levels of wisdom has implications again for how we think of wisdom in our ancestors, but not in a clear-cut fashion. Understanding that Aquinas thinks that levels of wisdom vary between humans means that although they may be included as intellectual and potentially wise beings, our ancestors might not be able to reach the same level of wisdom as us. It is possible that over time these levels fluctuated, progressing and regressing. In general, to the extent that our ancestors had a reduced capacity for interpreting sense-data, then they are likely to have had a reduced capacity for wisdom; they would come further down the spectrum. Of course, at some point the boundary mentioned above will be reached.

However, there are caveats to this. Why might this be? Aquinas is clear that as far as angels are concerned, the hierarchy of wisdom is set. The higher angels are wiser than the lower, and indeed contribute to their teaching. This rigidity does not necessarily continue among humans for two reasons. Firstly, there is greater division between the angels than there is between individual humans, with each occupying its own species. More importantly, Aquinas thinks that the way humans and angels progress toward wisdom is quite different. Angels attain wisdom in a single, immediate act. This is because they perceive truths naturally which we have to learn by progressing toward them.[57] The human path to wisdom is therefore slower and subject to error, and we vary in our levels of understanding.[58] So it is possible that the hierarchy of wisdom among humans is less clear—as I shall now show.

UNPREDICTABLE WISDOM

There are several reasons to suppose that the hierarchy of wisdom among humans is less static than it is among angels. Firstly, the discussion above says nothing of individual variation. We might be comfortable saying that early *Homo erectus*, with a brain size and development quite different from our own, could not become as wise as anatomically modern humans (if wisdom was a possibility for them at all).[59] However, I would be less sanguine about saying the same about Neandertals, for example. It seems quite possible that a sage Neandertal might be wiser than a foolish modern human.

Another caveat is that all intellectual souls have the same end-point or goal, which is contemplation of God.[60] This means that the hierarchy above breaks down in the beatific vision, in more ways than one. In the next life humans and angels are equal in beatitude, and Aquinas thinks that they will share a single society.[61] In discussing the angels, he does say that there are greater degrees of vision for some in beatitude.[62] However, this is not obviously a function of capacity for reason; so to the extent that any hierarchy of wisdom between different hominins can be identified, such a hierarchy may be upset if not entirely dismantled in the next life.

The reason for this potential upset is as follows: although wisdom is dependent upon an intellectual soul for its presence in a creature, the *extent* of that wisdom is not always proportional to our intellectual capacities. Aquinas recognizes various kinds or strands of wisdom, not all of which are determined by our natural capabilities.[63] Metaphysical or philosophical wisdom proceeds from natural principles, and those with greater capacity for rationality may be wiser in this way. But wisdom may also be given as a gift of the Holy Spirit, and the differing levels of wisdom that we may receive in this regard are not a result of our aptitude for reasoning but of charity, our love for God and our neighbor.[64] Charity is impeded in rational creatures by our tendency to do wrong and by the pull of other things on our energy and time.[65] So to say definitively that our ancestors were less wise than us requires not only saying that they possessed less rational capacity, which is perhaps plausible, but also that they were busier and morally worse people than we are. Neither busyness nor sinfulness seems to me obviously linked to rational capacity; I would be highly dubious of such a claim.

This returns to my earlier discussion of Aquinas' view that there can be physical restrictions on the expression of wisdom. I argued there that "intellect, while a good in itself, is not by itself a guide to the wisdom or overall perfection (or imperfection) of an individual." Here we see another thing needed to complete such a guide: the presence of charity. He does think that those with greater intellect will be able to progress further in philosophical wisdom, because this kind of wisdom is developed by our own activity and wise activity is a "highly intellectualised, argumentative, co-operative activity."[66] The gift of wisdom, though, is exercised through instinctive rather than reasoned understanding; moreover, this is a greater kind of wisdom because it more closely unites us to God.[67] Aquinas also discusses a third kind of wisdom, theological wisdom, which is knowledge delivered by holy contemplation and revelation.[68]

As discussed above, Aquinas thinks it possible for creatures with an intellectual soul to possess the gift of wisdom but be unable to act on it due to physical limitation of one kind or another. Now I have shown that he also thinks that the activity of this kind of wisdom is guided by instinct caused by charity. It is possible that by dint of increased cognitive ability modern humans have a head start on our ancestors in the pursuit of wisdom, since this will aid us in gaining philosophical wisdom. We could easily be overtaken, however, should it prove that they were kinder, holier, or more loving. Those who treat the pursuit of wisdom as exclusively intellectual will miss out. As long as they possessed the capacity for wisdom at all, our ancestors could well have been wiser than us—certainly on an individual basis, even if not on aggregate.

The last reason that the hierarchy of wisdom is not clear is that expression of wisdom is highly context-dependent. The most obvious outworking of this is in the virtue of prudence, or practical wisdom.[69] This closely related virtue represents the capacity to exercise good situational judgement—what Aquinas calls "right reason about things to be done."[70] The moral virtues—courage, generosity and so on—give us the desire to express them by being courageous, generous and so on. But it is not always clear how to do this. It is prudence that tells us what, in each context, being courageous or generous actually looks like.

It is not just practical wisdom, though, that is situationally sensitive. The same holds true of more speculative forms of wisdom. Since wisdom is an understanding of "first principles"—the ordering or foundational concepts of things—and since different practices have different principles, there is a sense in which there is a different kind of wisdom for each different practice. Aquinas calls this a distinction between wisdom in a genus and wisdom *simpliciter*: "Accordingly he that knows the highest cause in any particular genus, and by its means is able to judge and set in order all the things that belong to that genus, is said to be wise in that genus."[71] Knowledge of the highest cause of all things (i.e., knowledge of God) is wisdom *simpliciter*.

Given this distinction, it is possible for Aquinas to say that "Wisdom . . . can be said to have as many forms as there are legitimate theoretical sciences."[72] What this means is that just like prudence, speculative wisdom will look different depending on the context—although it will always have the ultimate purpose of understanding first causes. This is not a remarkable position for Aquinas; he is quite clear elsewhere that no human act can be understood apart from its circumstances and that right action for intellectual creatures takes a specific form for each individual.[73]

The point I want to make here is that our ancestors acted in circumstances profoundly different from our own. Any ascription of wisdom or its absence must therefore take this into account. It will not be enough to say that our ancestors were not wise because they did not do the wise things we do; nor, indeed, can we say that they *were* wise where we identify behavior that would today be classed as such. To make these kinds of determinations will require an understanding of the context in which they acted and the practices in which they engaged. Aquinas thinks that all humans have the same ultimate end and, in the broadest sense, the same needs and desires. So it is to be expected that there will be substantial shared ground between us. I am not suggesting that temporal and cultural distance poses an insurmountable obstacle to understanding. I am saying that understanding will not always be straightforward. There may be genera of wisdom in which they excelled and we do not, or vice versa; and where there are shared genera the exercise of wisdom in one case may look quite different from its exercise in another. It is to be expected that there will be varying degrees of wisdom between us and our ancestors. Understanding these differences in the form of a straightforward hierarchy, though, is on Aquinas' terms problematic.

CONCLUSION

In this chapter, I have used Aquinas to engage in an exploration of the theological consequences that follow an ascription of wisdom to our ancestors. I have concentrated on three distinct areas: firstly, what it means for a creature to possess the potential for wisdom; secondly, the significance of the transition from a creature that does not have the capacity for wisdom to one that does; thirdly, the way wisdom varies between different creatures. In the first, I showed that any corporeal creature with the possibility for wisdom must, on Aquinas' terms, be fully human. This means that any wise hominin is necessarily human (theologically speaking) and must be accounted for in Aquinas' "wisdom eschatology." In the second, I argued that this view does not entail a negative view of creation or of other humans. I also discussed the evolutionary process of becoming wise creatures, and how Aquinas' position entails quite different processes of change in the soul and the body to bring that about. In the third, I showed that levels of wisdom across species differ. This allows Aquinas to say that there is a hierarchy of wisdom. There are reasons to suppose that in general our ancestors might occupy a place below us, but this picture is complicated by the gift of wisdom, which depends on charity rather than intellect, and the situational variability of wisdom. For Aquinas, calling our ancestors wise is not merely an observation about particular capacities or behaviors. It means that we will share common ontological ground with them in a great many more ways, as embodied yet intellectual creatures—theologically speaking, human beings. Within this broader shared ground, though, it is not easy to locate ourselves. Aquinas' thought leaves open the possibility that the level and expression of our ancestors' wisdom may have been quite different from our own.

NOTES

1. Thomas Aquinas, *Summa Theologica,* trans. Fathers of the English Dominican Province (Notre Dame: Christian Classics, 1948), 1a2ae 57.2.

2. Norman Kretzmann, *The Metaphysics of Creation: Aquinas' Natural Theology in Summa Contra Gentiles II* (Oxford: Clarendon Press, 1999), 127.

3. Aquinas, *Summa Contra Gentiles,* trans. James F. Anderson et al., ed. Joseph Kenny, O.P. 5 vols. (New York: Hanover House, 1955–1957), 1.87.6; and Celia Deane-Drummond, *The Wisdom of the Liminal: Evolution and Other Animals in Human Becoming* (Grand Rapids, MI: Eerdmans, 2014), 69–70.

4. For a much more extensive account of Aquinas' views on these subjects, see Robert Pasnau, *Thomas Aquinas on Human Nature: A Philosophical Study of Summa Theologiae 1a 75–89* (Cambridge: Cambridge University Press, 2002).

5. Aquinas, *Disputed Questions on the Power of God,* trans. English Dominican Fathers, ed. Joseph Kenny O.P. (Westminster, MD: Newman Press, 1952), 7.4; *Disputed Questions on Truth,* trans. Robert W. Mulligan, S.J., James V. McGlynn, S.J., and Robert W. Schmidt, S.J., ed. Joseph Kenny O.P. (Chicago: Henry Regnery Company, 1952–1954), 23.6.

6. Aquinas, *Summa Contra Gentiles,* 2.58.3, 2.82; *Summa Theologica,* 1a 75.3.

7. Aquinas, *Summa Contra Gentiles,* 2.49; *Summa Theologica,* 1a 75.2.

8. Aquinas, *Summa Contra Gentiles,* 2.90–91; *Summa Theologica,* 1a 50.1; *Disputed Questions on Spiritual Creatures,* trans. Mary C. Fitzpatrick and John J. Wellmuth (Milwaukee: Marquette University Press, 1949), 2, 5.

9. Aquinas, *Summa Theologica,* 1a2ae 56.1.

10. Ibid., 1a2ae 57.2.
11. Aquinas, *Summa Contra Gentiles*, 2.90.
12. Ibid., 2.57.1; Pasnau, *Thomas Aquinas on Human Nature*, 25–46.
13. Aquinas, *Summa Theologica*, 1a 55.2.
14. Ibid., 1a 76.5.
15. Aquinas, *Summa Contra Gentiles*, 2.73.27, 2.90.2; *Summa Theologica*, 1a 76.5.
16. David Hershenov, "A Hylomorphic Account of Thought Experiments Concerning Personal Identity," *American Catholic Philosophical Quarterly* 82.3 (2008): 481–502.
17. Aquinas, *Summa Contra Gentiles*, 2.90.2.
18. Ibid., 2.70.
19. Matthew L. Lamb, "Wisdom Eschatology in Augustine and Aquinas," in *Aquinas the Augustinian*, ed. Michael Dauphinais, Barry David, and Matthew Levering (Washington, D.C.: Catholic University of America Press, 2007), 258–76; Matthew L. Lamb, "Eschatology of St Thomas Aquinas," in *Aquinas on Doctrine: A Critical Introduction*, ed. Thomas Weinandy, Daniel Keating, and John Yocum (New York: T & T Clark, 2004), 225–40.
20. Aquinas, *Disputed Questions on Spiritual Creatures*, 6.
21. Aquinas, *Summa Theologica*, 3a 49.3.
22. Ibid., 2a2ae 2.7, 3a 52.5.
23. Aquinas, *Summa Contra Gentiles*, 4.82.
24. David Clough, "The Anxiety of the Human Animal: Martin Luther on Non-Human Animals and Human Animality," in *Creaturely Theology: On God, Humans and Other Animals*, ed. Celia Deane-Drummond and David Clough (London: SCM Press, 2009), 41–60.
25. Aquinas, *Summa Contra Gentiles*, 2.95.2.
26. Aquinas, *Summa Theologica*, 1a 47.2.
27. Ibid.
28. Aquinas, *Summa Contra Gentiles*, 2.2.5.
29. Jame Schaefer, *Theological Foundations for Environmental Ethics: Reconstructing Patristic and Medieval Concepts* (Washington, D.C.: Georgetown University Press, 2009), 19.
30. Aquinas, *Summa Theologica*, 1a 85.7.
31. Ibid.
32. Ibid., 1a 76.2.
33. John Berkman, "Are Persons with Profound Intellectual Disabilities Sacramental Icons of Heavenly Life? Aquinas on Impairment," *Studies in Christian Ethics* 26.1 (2013): 83–96. See also Deane-Drummond, *Wisdom of the Liminal*, 68.
34. Berkman, "Persons with Profound Intellectual Disabilities."
35. Aquinas, *Summa Theologica*, 2a2ae 26.
36. Ibid., 1a 118.2.
37. Ibid.
38. Ibid., 1a 91.4.
39. Aquinas, *Summa Contra Gentiles*, 2.68.6.
40. Aquinas, *Disputed Questions on the Soul*, trans. John Patrick Rowan, ed. Joseph Kenny, O.P. (London: B. Herder Book Co., 1949), 7 Reply 1.
41. Aquinas, *Summa Contra Gentiles*, 2.89.6.
42. Aquinas, *Disputed Questions on the Soul*, 11, Reply 1.
43. Fabrizio Amerini, *Aquinas on the Beginning and End of Human Life*, trans. Mark Henniger (Cambridge, MA: Harvard University Press, 2013), 63.
44. Aquinas, *Quaestiones Quodlibetales*, trans. Sandra Edwards, Quodlibetal Questions 1 and 2 (Toronto: Pontifical Institute of Mediaeval Studies, 1983), 1.4.1.
45. Aquinas, *Summa Contra Gentiles*, 2.87.
46. Ibid., 2.49–50.
47. Aquinas, *Summa Theologica*, 3a 1–6.
48. Aquinas, *Summa Contra Gentiles*, 4.41.8–9.
49. Amerini, *Aquinas on the Beginning and End of Human Life*, 103–4.

50. Aquinas, *Summa Contra Gentiles,* 2.64.2.
51. Aquinas, *Summa Theologica,* 1a2ae 57.2.
52. Ibid., 1a 79.9.
53. Ibid., 1a 57.5.
54. Ibid., 1a 63.3.
55. Ibid., 1a2ae 2.8.
56. Aquinas, *Summa Contra Gentiles,* 1.91.10–11, 3.81; *Summa Theologica,* 1a 20.3–4.
57. Aquinas, *Summa Theologica,* 1a 58.3–4, 62.5.
58. Ibid., 1a 85.6–7.
59. Hélène Coqueugniot et al., "Early Brain Growth in Homo Erectus and Implications for Cognitive Ability," *Nature* 431.7006 (2004): 299–302.
60. Aquinas, *Summa Theologica,* 1a 62.1, 1a2ae 1.8.
61. Ibid., 1a 62.5, 108.8; *Disputed Questions on Spiritual Creatures,* 10.
62. Aquinas, *Summa Theologica,* 1a 62.9.
63. Paul Morrissey, "The Sapiential Dimension of Theology according to St. Thomas," *New Blackfriars* 93 (2010): 309–23.
64. Aquinas, *Summa Theologica,* 2a2ae 23, 45.4–5; Thomas S. Hibbs, "Wisdom Transformed by Love," *Where Wisdom Is Found,* ed. Robert B. Kruschwitz (Waco, TX: Center for Christian Ethics at Baylor University, 2009), 38–45.
65. Aquinas, *Disputed Questions on the Virtues,* trans. Ralph McInerny (South Bend, IN: St. Augustine's Press, 1999), 5.10.
66. Kretzmann, *Metaphysics of Creation,* 19.
67. Morrissey, "Sapiential Dimension"; Aquinas, *Summa Theologica,* 2a2ae 45.3.
68. Morrissey, "Sapiential Dimension."
69. Prudence is explored in greater depth in Celia Deane-Drummond, this volume.
70. Aquinas, *Summa Theologica,* 1a2ae 57.5.
71. Ibid., 2a2ae 45.
72. Gregory F. LaNave, "God, Creation, and the Possibility of Philosophical Wisdom: The Perspectives of Bonaventure and Aquinas," *Theological Studies* 69.4 (2008): 812–33.
73. Aquinas, *Summa Contra Gentiles,* 3.113; *Summa Theologica,* 1a2ae 18.3.

BIBLIOGRAPHY

Amerini, Fabrizio. *Aquinas on the Beginning and End of Human Life.* Translated by Mark Henniger. Cambridge, MA: Harvard University Press, 2013.

Aquinas, Thomas. *Disputed Questions on the Power of God.* Translated by English Dominican Fathers. Westminster, MD: Newman Press, 1952.

———. *Disputed Questions on the Soul.* Translated by John Patrick Rowan. Edited by Joseph Kenny, O.P. London: B. Herder Book Co., 1949.

———. *Disputed Questions on Spiritual Creatures.* Translated by Mary C. Fitzpatrick and John J. Wellmuth. Milwaukee, WI: Marquette University Press, 1949.

———. *Disputed Questions on Truth.* Translated by Robert W. Mulligan, S.J., James V. McGlynn, S.J. and Robert W. Schmidt, S.J. Chicago: Henry Regnery Company, 1952.

———. *Disputed Questions on the Virtues.* Translated by Ralph McInerny. South Bend, IN: St. Augustine's Press, 1999.

———. *Quaestiones Quodlibetales.* Translated by Sandra Edwards, Quodlibetal Questions 1 and 2. Toronto: Pontifical Institute of Mediaeval Studies, 1983.

———. *Summa Contra Gentiles.* Translated by Anton C. Pegis, James F. Anderson, Vernon J. Bourke, and Charles J. O'Neil. 5 vols. New York: Hanover House, 1955–1957.

———. *Summa Theologica.* Translated by Fathers of the English Dominican Province. Notre Dame, IN: Christian Classics, 1948.

Berkman, John. "Are Persons with Profound Intellectual Disabilities Sacramental Icons of Heavenly Life? Aquinas on Impairment," *Studies in Christian Ethics* 26.1 (2013): 83–96.

Clough, David. "The Anxiety of the Human Animal: Martin Luther on Non-Human Animals and Human Animality." In *Creaturely Theology: On God, Humans and Other Animals*, edited by Celia Deane-Drummond and David Clough, 41–60. London: SCM Press, 2009.

Coqueugniot, Hélène, et al. "Early Brain Growth in Homo Erectus and Implications for Cognitive Ability." *Nature* 431.7006 (2004): 299–302.

Deane-Drummond, Celia. *The Wisdom of the Liminal: Evolution and Other Animals in Human Becoming*. Grand Rapids, MI: Eerdmans, 2014.

Hibbs, Thomas S. "Wisdom Transformed by Love." *Where Wisdom Is Found*, edited by Robert B. Kruschwitz, 38–45. Waco, TX: Center for Christian Ethics at Baylor University, 2009. http://www.baylor.edu/content/services/document.php/79873.pdf.

Kretzmann, Norman. *The Metaphysics of Creation: Aquinas' Natural Theology in Summa Contra Gentiles II*. Oxford: Clarendon Press, 1999.

Lamb, Matthew L. "The Eschatology of St Thomas Aquinas." In *Aquinas on Doctrine: A Critical Introduction*, edited by Thomas Weinandy, Daniel Keating, and John Yocum, 225–40. New York: T & T Clark, 2004.

———. "Wisdom Eschatology in Augustine and Aquinas." In *Aquinas the Augustinian*, edited by Michael Dauphinais, Barry David, and Matthew Levering, 258–76. Washington, D.C.: Catholic University of America Press, 2007.

LaNave, Gregory F. "God, Creation, and the Possibility of Philosophical Wisdom: The Perspectives of Bonaventure and Aquinas." *Theological Studies* 69.4 (2008): 812–33.

Morrissey, Paul. "The Sapiential Dimension of Theology according to St. Thomas." *New Blackfriars* 93 (2010): 309–23.

Pasnau, Robert. *Thomas Aquinas on Human Nature: A Philosophical Study of Summa Theologiae 1a 75–89*. Cambridge: Cambridge University Press, 2002.

Schaefer, Jame. *Theological Foundations for Environmental Ethics: Reconstructing Patristic and Medieval Concepts*. Washington, D.C.: Georgetown University Press, 2009.

NINE

Practical Wisdom in the Making

A Theological Approach to Early Hominin Evolution in Conversation with Modern Jewish Philosophy

Celia Deane-Drummond

Where does wisdom come from and how can theological conceptions of human distinctiveness have anything to say to evolutionary anthropology? The background to this chapter is the philosophical notion of emergence, a view that has now become so common that it is often presupposed as a working paradigm in contemporary discussions related to science and religion, even though until relatively recently historically emergentist views of the early twentieth century were largely abandoned in favor of a bottom-up scientific positivism.[1] Nearly thirty years ago Philip Clayton claimed that emergence is "a conceptually coherent and empirically viable middle ground between dualism and reductionism."[2] As applied to living systems, emergence in general refers to top-down causation and the impossibility of reducing the complex function of the whole to the constituent parts. The cell, the organism, and even consciousness itself can be viewed through the lens of emergence. Ian Barbour contends that emergence is "the claim that in evolutionary history and in the development of the individual organism, there occur forms of order and levels of activity that are genuinely new and qualitatively different. A stronger version of emergence is the thesis that the events at higher levels are not determined by events at lower levels and are themselves causally effective."[3] Such claims for emergence thus far might either appear innocuous or seem to say not much more than that the whole is greater than the sum of the parts. Most scholars who are in favor of emergence resist a strong or radical emergence that claims that something arises whose nature and existence are not constrained by the constituent parts.

Authors such as Paul Davies, citing Harold Morowitz, press harder and claim emergence provides a *metaphysical* worldview where "the transcendence of God is itself an emergent reality."[4] On one level, theists may be drawn to stronger versions of emergence insofar as they seem to support a progressive trend toward complexity in evolutionary processes that could be viewed as in some way compatible with teleology and divine purposefulness. But I suggest that emergence in human evolution is insufficient and may even be just as misleading as an alternative classic neo-Darwinian view, since it can imply a progressive view of human becoming that marches ahead, leaving behind earlier, less complex forms.[5] Many theologians drawing on the idea of emergence forget the historical background to evolutionary theory in the work of James Hutton, who, in a fascinating way, introduced the idea of *cyclical time* to the deep history of the earth, believing that the beginning and end could not be accessed by scientific inquiry.[6] Emergence is characteristically nonlinear, but still progressive, and analogous to Simon Conway Morris' discussion of the convergence theory of morphological forms.[7]

How far do emergence theories cohere with the exploration of an evolution of human wisdom? While in broad terms emergence is logically correct, in that more complex forms could be viewed as those that *emerge* from the least complex, I will show that it is the *metaphysical* appropriation of emergence and its adoption, at least in some cases, as a linear arrow of time that is more problematic, both theologically and in evolutionary anthropology. At the same time, an alternative highly contingent form of Darwinianism, illustrated by the work of Stephen Jay Gould, in which complexity emerges as a diffusion effect as a result of a random walk, seems too thin, and, as Paul Davies points out, "the rise of complexity would be rather trivial."[8] Hence, one of the difficulties arising from a discussion of emergence is navigating not simply the alternatives of reductionism or dualism, but stronger or weaker versions of teleology, sometimes referred to as teleonomy by biologists anxious to reject "purposeful" evolution. For example, is Simon Conway Morris' contention that human beings are "inevitable"[9] far too strong a statement given what can be uncovered about early human evolution? At the same time, is the opposite extreme, that humanity as we know it today emerged from a random walk from prior basally complex forms, also too simplistic? Can a theological analysis provide new insights and questions that need to be asked in developing theories about human evolution?

Practical wisdom, as a complex intellectual virtue composed of different component parts, provides, it seems to me, not only a good case study for thinking through emergentist theories, but also a way of uncovering important aspects of the overall process of human evolution. In order to develop this argument, I will touch on some new research on the evolution of social complexity in early hominins that will be developed in more detail in the anthropological chapters in this volume. I suggest that Alan Mittleman's discussion of Jewish interpretations of early human evolution is particularly interesting, since some of these interpretations, at least, seem to reflect a deep memory that is resonant with what is being uncovered in this research. I will then develop the argument further by analysis of the specific component parts of practical wisdom that seem to be the most relevant to this discussion, especially those distinctive human abilities to look ahead, *foresight*, and look back, *memoria*. While my main interlocutor here will be Thomas Aquinas, Joseph Soloveitchik will be useful for my discussion of foresight, in conversation with contemporary research by neuroscientists such as Martin Seligman, while Hans Jonas will be useful for filling out my discussion of *memoria*.

TRADITIONS OF HUMAN EVOLUTION

For evolutionary anthropologists, the human capacity for *symbolic* thought is confined to anatomically modern humans; thus symbol becomes the mark of what makes humans distinct.[10] But planning for the future and the ability to conduct long-range exchange of goods shows that at least by the Pleistocene the ability to show *foresight* must have been intact. Using stone tool assembly, art, and non-ritual activity, including deliberate burial practices, Philip Chase and Harold Dibble suggested that Neanderthals in the Middle Paleolithic period possessed foresight, which they described as the ability to create something from a mental image.[11] Their work was later criticized regarding assumptions made about what Neanderthals may or may not have intended, since there could have been other environmental and geological explanations for what they had uncovered.[12] Of course, such explanations were equally speculative, while still leaving open the possibility of intentionality.[13] Sarah Tarlow, like many others since, has questioned our current ability to recognize distinct biological and cultural aspects of hominin emotional life, arguing that there is a strong tendency to read into that record what is perceived in our own culture, thus providing a form of legitimization.[14] A measure of caution, and perhaps humility, about the limitations of understanding and the need for a qualified hermeneutic is therefore appropriate.

The thesis that symbolic thinking defined or tracked the emergence of anatomically modern humans has begun to be challenged by cumulative evidence of complex artifacts existing in association with very much earlier hominin remains, including the use of ocher, stone tools, blades, and so on. These glimpses of some sort of symbolic capability in the material record imply a far deeper history to such symbolic capacities, even if their fullest development came much later.[15] In a fascinating way, there is also growing cumulative evidence that different material records of such specialist capacities first come and then disappear at a range of geographical sites.[16] Is the development of advanced cognitive abilities, including what could be termed the development of practical and social intelligence, a much more sporadic process that included *flickerings* rather than a single unified process, as has been described for complex cognition more generally?[17] These results are consistent with the thesis that the evolutionary lineage *Homo* was a slowly evolving and even sporadic community niche, certainly not some kind of progressive or even directly linear emergence. This complex dynamic system included cognitive, social, and ecological components interacting with each other and with the genotype in complex feedback loops.[18]

The challenge with any results such as these is how to interpret such findings theologically. Alan Mittleman provides a fascinating argument for the reasons why, when it comes to scientific debates about human nature, ancient Jewish scholarship needs to be taken seriously.[19] Mittleman's philosophical scrutiny of the implications of evolutionary and biological accounts of human beings faces the challenge that all religious theists face, namely, how to make Jewish religious accounts seem *reasonable* for those in or outside that tradition and not just assertions of faith based on divine authority arising from a lived tradition.[20] Now, while I agree that the concept of transcendence comes to the surface in places other than religion, and that resorting to a religious "trump card"[21] as he calls it is not all that helpful, many theologians, including myself, would be prepared to go rather further in giving weight to religious experience as well as what counts as philosophically reasonable. He leans toward such a view himself by stressing the way Jewish ideas are at work in the moral and religious life, so "even stringently secular projects such as Kant's work off the normative capital of generations of biblically informed moral imagination, piety and commitment."[22]

Yet, it is his interpretation of image bearing that is of direct relevance to the present discussion. For him, the Genesis accounts have a bearing on how values have come into existence and are in a primary sense about "responsibility, respect, and recognition of the individual worth of persons."[23] At the same time, human beings are *intentional* rather than accidental creations, and did not simply emerge, but arose from a special act of God.[24] But the most interesting part is that "what is unmistakably clear is that God takes a moment for deliberation. He does not just command. He considers. The human being is a product of reflection and choice."[25] Recursive thought is distinctly human, and is also in a fascinating way attributed to God in the making of human beings. Yet, importantly, "human beings are at best contingently not categorically superior to other beings," and "Jewish sources express a remarkable ambivalence about the human, as if to fold humanity back into the world of all other creatures."[26] The divine image and likeness, at least in rabbinic texts, meant initially physical resemblance, and some texts claim that angels mistook Adam for God.[27] The form Adam shares with God is not so much physical as energetic and radiant, though after Adam's sin the only possibility is taking responsibility by acting like God, especially in humanity's mandate to rule over animals, *imitatio Dei.*[28]

The indecisive beginning to human origins is also reflected further in a rabbinic interpretation of "us"—in "let us create man in our image, after our likeness" (Genesis 1:26)—in terms of God's angelic hosts, who remained divided. Some angels argued for and some against the creation of human beings. Those against the creation of humanity were destroyed, but human sin has since proven that angelic skepticism was correct. The anxiety in rabbinic sources about the worth of human beings relates to their perception of human moral weakness, rather than implying that their existence is pointless. Rabbinic ambivalence about whether humans should have been created at all is mirrored in a portrayal of ambivalence in God. If humanity is created, evil will arise, but if not, then there will be no righteousness.[29] In the end, God lets compassion overrule judgment and creates humanity in spite of their conflicted and potentially sinful nature. The propensity for evil rather than good is also, according to the rabbinic traditions, a divinely created force.[30] This direct divine creation of the propensity for evil is important for arguments on human distinctiveness, since it is not simply a question of humans emerging from their untamed and supposedly "sinful" animal-like tendencies, but much more complicated. The worthiness of humans hangs in the balance: if they act in the image of God, then they will rule in right relationship over animals; if not, they will become just "a middling species among them."[31]

Although Mittleman is correct that the primary intention of these texts is to speak about the value of humanity and human moral existence, I find the indecision on God's part on the basis of that moral ambiguity really interesting. Might this account of God's vacillation reflect a distant memory of early human origins whose coming into being in evolutionary terms was both tenuous and to some extent highly contingent, to say the least? And why is the rule over animals and human relationship with them highlighted if this early Genesis text was, according to Mittleman, merely a commentary on human values? In other words, are there deep memories embedded here which speak of the close and yet highly ambivalent relationships of humans with other creatures? While literalism of any crude sort needs to be eschewed, there are intriguing analogical resonances here with evolutionary anthropology. Are there other ways of considering the specific and distinctive abilities of humans that gradually show up as artifacts in the evolutionary record? More specifically, are there any specific clues about practical wisdom that could be brought to bear on the more scientific evolutionary anthropological studies of the evolution of wisdom?

MAPPING PRACTICAL WISDOM

I suggest that the medieval scholar Thomas Aquinas is a helpful resource in the Christian tradition that can be brought to bear on these difficult questions. It is my contention that the first form of wisdom that evolved in the human community was likely to be practical, rather than the intellectual virtue of wisdom proper, which was cognizant in its relationship with God and in all probability demanded speech. Although complete practical wisdom involves deliberation, judgment, and then action, might there have been, at least initially in early hominin evolution, non-linguistic forms of deliberation through other non-linguistic forms of signaling? Or perhaps, in addition, the different elements necessary for that complete form of practical wisdom did not emerge all at once, but some show up before others in evolutionary terms? That would be consistent with what is generally known about the evolution of other complex characteristics, namely, that different aspects are co-opted and exist in a complex relationship to the observed characteristics. Further, the biological basis of such intellectual capabilities needs to be distinguished from its cultural expression, while recognizing that the relationship between them and the boundary of nature-cultures is far more blurred in comparison with the standard nature/nurture rhetoric. So, the characteristics of practical wisdom described below should not be thought of as specific or isolated "traits" that are selected for independently of each other, but rather part of a gradual building up of social intelligence and mature imagination. Bearing this in mind, a Thomistic analysis of practical wisdom starts to lay out the potential parameters for further discussion.

Within practical wisdom there are various components that could usefully be parsed out. Practical wisdom, or prudence, for Thomas, has eight components, namely *memoria* (memory), *intellectus* (insight/intelligence), *docilitas* (teachableness), *solertia* (acumen), *ratio* (reasoned judgment), *providentia* (foresight/prevision), *circumspectio* (circumspection), and *cautio* (caution).[32] Cognitive prudence, consisting of the first three—*memoria, docilitas,* and *solertia*—is relevant as these three elements show the kind of mindset that is necessary in order to make wise decisions. *Memoria* is a memory that is "true to being," which means that "it contains in itself real things and events as they really are and were."[33] Having an *accurate* memory of prior experiences would be particularly essential for early human survival in the Pleistocene, especially since the ability to record memories in written form was not yet possible. *Docilitas* is essential for passing skills from one person to the next. But it is also a reminder that complex decisions are never simply just about an individual's position in detachment from the views of others, and would have been an important part of forming communities. *Solertia* (acumen), which includes the ability to act clearly and well in the face of the unexpected, was also essential for early humans that were still threatened by the presence of large predators. The ability to show *solertia* was a matter of survival.[34]

Insight (*intellectus*) and reasoned judgment (*ratio*) in the list of components of prudence related to cognition needed to be brought to bear, in that incorrect judgments could have led to fatal consequences. *Foresight* (*providentia*) is characteristically human in a way that *memoria* and *docilitas* are not, even though there are specifically human ways of remembering that are not possible for other animals. Insofar as *providentia* is linked with divine Providence, it bears something of the character of the divine in us. The judgments of practical wisdom are not fixed or certain in ways that might be the case if it were simply an application of rules or principles. This is important, since it shows that at least elements of practical wisdom could arise before any fixed rules or specific codes of conduct were developed. Yes, there must have been a sense of what constituted the good of the community, but rules for that community would not necessarily emerge in a clear-cut way immediately. Rather, for Thomas, "because the subject matter of prudence is composed of contingent individual incidents, which form the setting for human acts, the certitude of prudence is not such as to remove entirely all uneasiness of mind."[35] Thomas also includes *circumspection* and *caution* in the list of those components of prudence that are concerned with putting knowledge into action. Circumspection is the ability to judge situations clearly and well and gather all the relevant information. A correctly informed caution as a component of prudence gives good judgment about refraining from action where appropriate and recognizing risks.

HOMO PROSPECTUS?

Of all these parts, foresight is, it seems to me, one of the most interesting to consider in more detail, since it is the one in which particular human goods can be set. According to this view, human distinctiveness is less about overcoming sinful or animalistic tendencies and more about developing a specific capacity to look ahead and imagine the future in realistic and profound ways. Indeed, Thomas believed that foresight was the most significant and the dominant part of practical wisdom, since without it good judgments of prudence were impossible.[36] Even the classical definition of prudence, *prudens*, comes from the phrase "looking ahead," *porro videns*, and for Thomas, believing that prudence is a cognitive rather than affective power, "the prudent man is keen sighted and foresees how uncertainties will fall."[37] Foresight, which he also called human providence, was strictly in practical matters about the future, rather than the present or past. For the past and present involve a kind of common sense necessity or inevitability since past events cannot be undone, so "consequentially only those future contingents which a man can shape to the purpose of human life are matters for prudence."[38] The combination of contingency and purposefulness is what practical wisdom is all about. Thomas concludes, therefore, that in the midst of all the different parts of prudence, there is one that is "the chief to which the others are subordinate" and "which is significant and dominant, and which gives unity to the whole."[39] And that one component is *prevision* or, as used in other texts, *foresight*. Further, that prevision is practical rather than purely theoretical, so it is less about universal truths than the ability to make effective practical decisions where there are alternative possibilities. Thomas also recognized that it is possible to be too preoccupied about the future in a way that is a distraction from present tasks, a capacity that he names as distorted solicitude acting in concert with foresight. So "duly looking to the future belongs to prudence, yet such forethought and concern would be inordinate were a person to make temporal things, to which the terms 'past' and 'present' apply, his goal in life, or were he to seek superfluities beyond the needs of his present life or to anticipate the fitting time for solicitude."[40] Such a capacity to judge appropriately between good ends would have been critical for the survival of early humans, who were up against threats of being eaten by predators and a host of other environmental challenges. This solicitude has to be at the right time in the right place and for the right reason, and when it is not, or if it conflicts with a perceived God-given task, it is part of what Thomas calls "sham prudence." Thus the long reach of prudence has a transcendental goal even if more immediate decisions are focused on practical demands.

Martin Seligman argues that "we are misnamed" and modern humans should not be called wise at all for the simple reason that none of us really is wise, and thus it is an aspiration only. His suggestion is that unlike other social animals, "the unrivaled human ability to be guided by imagining alternatives stretching into the future—'prospection'—uniquely describes *Homo sapiens*."[41] Peter Railton says much the same in describing wisdom as the "outcome" of human nature, rather than its origin. He adds that "much of what we learn over the course of our lives arises not from solo experience, but from observation and interaction with others," hence, expanding *Homo sapiens* into *Homo prospectus socialis*.[42] Seligman and Railton accept that this ability is integral to the slow learning of wisdom. The difference, compared with their approach, is that I am prepared to delineate the prospective and socially intelligent characteristics of human beings as being *integral to what practical wisdom means*. It may not be fully developed, but it is still present. It is also worth considering the evidential basis for human prospection that I prefer to call foresight. For Seligman the designation of *Homo prospectus* amounts to a new turn in psychological research that has, in the past, dwelled on an image of human agency as locked into its past history. Inasmuch as he argues that human distinctiveness is best parsed out in terms of how humanity is drawn into the future this makes sense. However, I will argue that the crucial capacity for memory understood in terms of *memoria* should not be forgotten or dispensed with in the enthusiasm to think through the benefits of imagining human becoming in terms of future scenarios. For medievalists, practical wisdom was represented with three eyes, looking to the past, present, and future.[43] Anticipation is a characteristic of social animals; what is much more difficult is the ability to plan ahead with a good chance of successful prediction in order to regulate present behavior, and for that, humans need "a very expensive brain and a very long apprenticeship in life."[44]

Thomas Aquinas, in common with contemporary psychology, recognized the importance of learning and memory as providing the background into which decisions about the future are made.[45] This latest discussion of *Homo prospectus* puts perhaps rather more emphasis on the importance of *affective* as well as cognitive qualities and *intuition* as well as deliberation in prospective decision making.[46] However, even the Thomistic account recognized the importance of the theological virtue of charity, and there is a close relationship between practical wisdom and charity. So, "now the initial activity of the appetitive power is loving, as we have seen. It is in this sense that prudence is said to be love, not that of its nature it is a kind of love, but because its activity is caused by love."[47] Further, he allows for insight as a critical component of practical wisdom, which, broadly speaking, seems very close to the way psychologists are presenting intuition. For Thomas Aquinas, insight is integral to the ability to have knowledge. Knowledge of the past is related to memory, while knowledge of the present—either certain, as a matter of necessity or contingent—he brackets as both insight and intelligence.[48] This intelligence/intuition is placed in a different category from reasoned judgment.[49] That this is intuitive knowledge is actually spelled out by Thomas when he claims that "Cicero places intelligence and Macrobius places intuitive understanding—which amounts to the same—as part of prudence."[50] The role of intuition in prudence seems to be, for Thomas at least, "a certain correct appreciation of an ultimate principle assumed as self-evident."[51] Those principles are general principles that are true in practice, such as *do no evil*; but significantly, it is also the way that ultimate principle is expressed practically in the concrete situations to hand: "a certain correct appreciation of a particular end."[52] But what is really fascinating is that Thomas seems to place insight even before the slower task of deliberative reasoning. So "those beings where the power of understanding or intelligence flourishes in full vigour have no need for reasoning, for they grasp truth by simple insight, as do God and the angels."[53] But if that is the case, then, theoretically at least, it should be quite possible for early hominins to arrive by intuitive intelligence at judgments of prudence without necessarily using a more complicated deliberative form of reasoning. Peter Railton, commenting on more recent psychological research, notes that "the rapidity of intuitive responses need not mean that they do not involve complex processing of information or calculation of expected value and risk."[54] A crucial distinction between Thomas and the contemporary discussion as laid out by Seligman, Railton, and their colleagues is that for Thomas, following Augustine, "prudence is the love which well discerns between the helps and the hindrances in our striving toward God."[55] Hence, there may be an immediate further end in prudence but also a final end. In evolutionary terms that striving toward God presupposes that some awareness of God is present in the earliest human consciousness and that a full conception of God is always out of reach. Just as it is almost impossible to track accurately subtle changes in human mental states, so religious consciousness is equally elusive. Nonetheless, it is possible to speculate that the ability to prospect into the future could just as easily have been *fostered* by religious experience, rather than necessarily assuming that such abilities were a prelude to religious consciousness. Roy Bauminster, by comparison, uncritically adopts Norenzayan's *Big Gods* thesis.[56] Prudence is situated between intellectual and moral virtues rather like a compass; hence, it attunes the prospecting mind to good ends. While the classic tradition represented by Thomas Aquinas shows that he was aware of such capacities through the ability to have foresight, Seligman's *Homo prospectus* lacks any prospecting within a wider consideration of the context of those good purposes as under the general direction of divine Providence. That divine Providence need not be rigid or deterministic, since divine Providence still protects contingency for those things that are intended to be contingent. So "the effect of divine Providence is for a thing to come about not anyhow but in its own proper style, necessarily or contingently."[57]

Jewish rabbi and philosopher Joseph Soloveitchik in his book *The Emergence of Ethical Man* describes human distinctiveness in terms of morality.[58] It seems that Soloveitchik also anticipated Seligman's new research agenda in psychology. Thus, for Soloveitchik, it is this capacity to *look ahead* that marks out human distinctiveness, so "when man anticipates himself in prospect, he chances upon something unique—he has the peculiar ability to encounter nature and contrast himself with it in terms of I and it."[59] Like Seligman and his associates, Soloveitchik believes there is "nothing mystical" in such cognitive drives at the dawn of human *noesis,* that is, mental perception and intelligence.[60] It is in looking back, argues Soloveitchik, that humanity experiences a sense of unity with "Mother Nature," but with the ability to prospect comes an experience of otherness from that nature. The experience of approach to another subjective existence, I-Thou, is, however, different. In this case, the experience of retrospect brings a separate existence into view, while prospect allows for the possibility of union and intimacy.[61] Soloveitchik then concludes that the ethical task demands a sense of a common future and destiny exclusively with other human beings. But what if, as seems likely, early humans perceived other animals as subjective in the manner that

Soloveitchik seems to insist only applies to other humans? Soloveitchik anticipates the prospective mind, but contemporary anthropological studies resist the idea that alienation from nature was a prerequisite to building human moral communities, or that prospection is necessarily premised on imaging ourselves as sharply separate from other creatures.

MEMORIA AND PRACTICAL WISDOM

The second aspect of Thomas Aquinas' work on prudence that is worth paying particular attention to is his discussion of *memoria*. He lists this characteristic first in the sequence of components, and for good reason, since without *memoria* prudential reasoning would be impossible. There are four aspects of memory that Thomas highlights.[62] The first is related to association with unusual images, and these bodily images help with retention. Second, memory requires an ordering of information in ways that make it easy to move from one thing to the next. Third, there is effort in remembering, so that we are more likely to remember if we take the trouble to keep the memory of images imprinted on the mind. Fourth, returning frequently to the things that we want to remember helps them to stay in the memory, so frequent use entails, as Aristotle suggested, customs that become second nature. Memory, therefore, comes through diligent practice, but it is also possible that memories come to mind through the work of grace. So "the things we often think about are quickly recalled by a sort of instinctive process."[63] Memory should not be thought of in Thomistic terms as a virtue related to prudence as a species is to a genus, but rather, "one of the conditions required for prudence is good memory, so that, in a sense, it is an integral part of prudence."[64] Further, it is the comparative quality of human memory that marks out its distinctiveness, so that "as to memory, man not only has it the way other animals do, but also in the form of reminiscence, a quasi-syllogistic search among memories of things past in their individuality."[65] Accordingly, the searching out of different possibilities in foresight is in some senses parallel to the ability to search among memories. Other animals use their estimative sense, while humans use what Thomas called the "particular reason" in distinguishing memory, but he concludes that in humans both cogitation and memory "reach so high" "by a kind of overflow," that "they are not new powers, but the same powers, more perfect than they are in other animals."[66] Further, while the distinctly human imagination is a prerequisite for the more elevated powers of memory and foresight, it is not identical to it. Thomas also cites Augustine, who claimed that there are three kinds of vision, a corporeal one in the external senses, a spiritual one in the imagination, and an intellectual one in the understanding.[67] To some extent, prudence taps into all three, in that practical wisdom can be received by grace as well as acquired through understanding.

In order to appreciate the full significance of *memoria* for the present discussion, it is worth elaborating the perception of the creative and positive aspects of memory in the medieval context. Medieval historians Mary Carruthers and Jan M. Ziolkowski summarize this perspective as follows:

> [A]s an art, memory was most importantly associated in the Middle Ages with composition, not simply with retention. Medieval *memoria* took the inventive function of human memory for granted, and emphasized it. Indeed, those who practiced the crafts of memory used them—as all crafts are used—to make new things: prayers, meditations, sermons, pictures, hymns, stories and poems. Students of art and literature have long remarked on the intensely pictorial and affective qualities of these arts in the Middle Ages.[68]

The point is that while eventually texts and words came to have pride of place for the development of memory in medieval contexts, the creative use of imaginative images in memory has parallels with the creative use of imagination in foresight in a way that is neglected in current psychological accounts. Further, it is the specific human capacity for recollection that is important, so the process of remembering as recollection is active, unlike the more passive, impressionistic use of memory in other animals.[69] Recollection is developed through habitual use, and in association with specific images, or, in linguistic contexts, words.[70] The ultimate goal of remembering for the medievalists was a religious one, the meditative "memory of God."[71] Further, Thomas Aquinas was ready to admit that "some animals have a kind of prudence," and that this prudence works through "sense perceptions" including "not only the ability to sense in the present, but also a memory of the past."[72] The difference between humans and other animals has to do with a deficiency in memory in the latter, who, unlike humans, are unable to recollect.[73] And "recollection is a kind of inquiry, because the person recollecting does not proceed by chance from one thing to another, but he proceeds with the intention of coming upon the memory of some thing," and is found in those with "a natural power of deliberation," which is also achieved through reasoning.[74] But there is a difference between the kind of deliberation that happens in memory and that involved in intellectual processes, since the person recollecting does not have the power to stop the process.[75] For that reason, Thomas, following Aristotle, thought the memory was in a "physical organ," even though it connects with the intellect.

Now, in the context of human evolutionary history, it seems quite likely that vivid images (that are even translated sometimes as *phantasms* in Thomas' text) started to be retained in the memory as a first step toward the development of foresight. Things and striking visceral images are much more memorable than words, so the ability to recollect in this way does not have to be reliant on explicit linguistic tools, even though that assists in communication between persons.

PICTORIAL HUMANITY

Further light on the potential role of a distinctly human memory is shed through analysis of Hans Jonas' *Phenomenon of Life*. Jonas begins with the capacity for a distinctly human imagination in its perceptive ability to distinguish between observing an object and the creation of an image of that object through what he calls "pictorial man."[76] Jonas recognizes that this capability is distinct, but that there is no immediate usefulness for such a characteristic; "an image-making creature, therefore, is one that indulges in the making of useless objects, or has ends in addition to the biological ones, or can serve the latter in ways remote from the direct usefulness of the instrumental things."[77] This is a new relationship with objects from instrumental to seemingly useless purposes. But is this seeming uselessness a correct interpretation? If image making is integral to memory and foresight, then these are important and arguably essential ingredients of the capacity for wisdom. This follows directly from the point I raised concerning the importance of image making in memory.

As Jonas is well aware, image creation is not the same as straight *imitation,* which would be the creation of a fake; rather, image making involves a deliberate artistic *choice* of which aspects to emphasize. While, for memory, it may be harder to control those choices in recollection, this is not the case for foresight. Language is a specific form of image making, but does not need to be coincident with it. The first requirement in any image making is the "ability to perceive likeness," and not only to perceive it, but to perceive it in a certain way. A bird, mistaking a scarecrow for a man, does not register an intermediate state, but is either deceived or not; thus "when the deception breaks down, only straw, sticks and rags remain. For the animal, similitude does not exist."[78] In other words, this is not just related to perception, but perception as "mere likeness," and image showing a characteristic range of greater or lesser likenesses. Pushing this further, Jonas argues that the principle involved is the separation of form from matter, which is a non-perceptual quality of distinction. And a prerequisite of this ability is the ability to step back from the environment and to distinguish appearance from reality. So "this free possession is first achieved in the internal exercise of imagination, by which, to the best of our knowledge, human memory is distinguished from animal recollection."[79] What is interesting here is that he uses *recollection* to describe the more limited capabilities in other animals, but *memory* for humans. The important difference is the exercise of imagination, so for animals, "there is nothing to show this kind of remembering enjoys an imaginary presence of its objects, and everything to argue against the assumption that, if it does, this presence is at the subject's command, to be summoned and dismissed at will."[80] Here he parts company with Thomas in giving rather more weight to the human ability to conjure up particular memories through a deliberate process. For Jonas, human memory "has images of things at its call."[81] For Jonas, the elemental power of the imagination is primary, so "imagination separates the remembered eidos from the occurrence of individual encounter with it, freeing its possession from the accidents of space and time. The freedom so gained—to ponder things in imagination—is one of distance and control at once."[82] What follows is a movement from internal imagination to external image that again becomes an object of perception, but in a new way, with recognition of distance from experience. Such memories are then "shared" and become common property. But, in a crucial move, Jonas makes the following point: "the remaker of things is potentially also the maker of new things, and the one power is not different from the other."[83] But the translation of such capabilities into external existence also shows a physical aspect of this inner change. It might show up as the production of an image, or a thing or any other physical action, all of which are governed by "image." Jonas notes that this could take the form of writing, but also dance, and use of the hand in "motor translation of form" "as the condition of all technology."[84] Such eidetic control of mobility arises from a freely chosen expression of form and complements the eidetic control of the imagination, thus linking up *homo faber* with *homo sapiens* through *homo pictor*. Jonas does not elaborate how such a dynamic might function in memory and foresight, or how such activities are also an expression of wisdom, but these are precisely the components that are necessary for its proper function. Yet he does insist that this transition to attempted sensible likeness is sufficient for the category of the human, and amounts to a "metaphysical gap." This gives some insight into how *memoria* might develop in prudential reasoning, and the sharing of simple artifacts in early hominin societies is one way in which such memories could be extended across different geographical regions.

CONCLUDING REMARKS

My argument in this chapter is that a distinctly human wisdom did not surface overnight nor follow a linear process of emergence according to a direct arrow of time. The fact that what look like specific capacities to create artifacts, which seem to require the kind of imagination Jonas would attest as metaphysically distinct from that in other animals, first show up in the deep archeological record, but then disappear either demonstrates the contingency of the process of human becoming wise, or attests to the difficulty of finding consistent archaic records, or some combination of both. Early Jewish scholarship elaborated by Alan Mittleman hints at the possibility of a deep memory of uncertainty in the creation of human beings. Practical wisdom's capability of foresight is one of the most striking features of human communities, and contemporary psychologists such as Martin Seligman even resist naming human beings as wise due to their (mistaken) belief that prospecting offers a different kind of mental activity. I argue, by contrast, drawing on Thomistic sources, that *Homo sapiens* as *sapiens* necessarily includes foresight. Further, I contend that foresight is the work of an active imagination that includes an awareness of different possibilities. But it seems more likely that the companion ability *to remember* in a specifically human way was just as important as the imaginative ability for planning and foresight. Early medieval sources, including Thomistic texts, attest to a different way of perceiving memory, less as a recall and more as an imaginative way of seeing the past according to specific images. Such pictorial learning resonates with Hans Jonas' discussion of pictorial humanity. Indeed, the imaginative ability of humans to transform an object into an image of that object that Jonas calls *form* is likely to mark a significant step in human becoming. Important though this might be, I am less inclined, compared with Jonas, or Seligman for that matter, to identify one salient feature as *the Rubicon* in human cognitive and moral evolution. Human becoming wise is a collective endeavor, rather than an individual one, and it is one where a number of different features or components of that wisdom eventually converged.

How might the arguments in this chapter fit with contemporary debates on the evolution of morality? Moral psychology has focused on type 1 emotional, quick, intuitive judgments that have then been compared with the slower and deliberate judgments of type 2 which involve more deliberative reasoning. As Michael Tomasello suggests, moral psychology has not attended sufficiently to how human beings "understand their social worlds rationally and make rational decisions based on this understanding."[85] And yet the elements of practical wisdom identified in this chapter involve a combination of both psychological elements: the intuitive imagination *and* the ability to recollect past events, reason, and predict future outcomes accurately. A parallel discussion on the evolution of empathy, which has been given far more focus in the moral psychology and evolutionary literature compared with wisdom, indicates that compassion involves cognitive processes as well as emotional ones.[86] Further, perhaps it is worth stressing here that, like compassion, in an evolutionary sense practical wisdom necessarily emerges in collective and not just individual contexts. The philosophical literature to date finds it hard to incorporate such a collective perspective, but it is crucially important in the present context of this volume on the evolution of wisdom. There would be little advantage for human communities if a single individual had abilities of foresight, or even keen patterns of remembering that could not be shared; rather, the capacity for interdependence and joint collective attention forms the background in which such capabilities should be understood, at least in evolutionary terms.[87] Given this background, technological practices that indicated either symbolic meaning or were non-utilitarian go back very deep into human evolutionary history.[88] Such iconic and symbolic social practices would have reinforced and even provided the basis for imaginative and shared memories in a positive feedback loop that eventually provided the basis for a shared sense of cooperative rationality in humans. As human communities searched for a transcendent Other, an experience of that Other gradually dawned, filling out memories and eventually providing foresight with a sense of divine Providence. That expression in the form of Monotheism and a loving, caring, Providential Creator who had oversight of the course of human affairs would come very much later in human history. But in retrospect it is possible to interpret the yearnings of the very earliest humans theologically as a slow reaching out toward the transcendent, the inklings of which were in place long before our own identified species, *Homo sapiens sapiens*, first walked on the savannah hundreds of thousands of years ago. The question of whether such hominins were made in the image of God or not is largely irrelevant to this discussion, but if image bearing at least in a minimal sense requires self-awareness of God and faithful response to God in terms of action and practice, then the evolutionary emergence of such capacities did not take place overnight, but through a long and patient process of human becoming and evolutionary change.

NOTES

1. As discussed in Nancey Murphy, "Introduction," in *Evolution and Emergence: Systems, Organisms, Persons*, ed. Nancey Murphy and William R. Stoeger (Oxford: Oxford University Press, 2007).

2. Philip Clayton, *God and Contemporary Science* (Edinburgh: Edinburgh University Press, 1997), 248.

3. Ian Barbour, "Neuroscience, Artificial Intelligence and Human Nature: Theological and Philosophical Reflections," in *Neuroscience and the Person: Scientific Perspectives on Divine Action*, ed. Robert J. Russell, Nancey Murphy, Theo C. Meyering, and Michael A. Arbib (Notre Dame: University of Notre Dame Press, 1999), 271.

4. Paul Davies, "Introduction: Toward an Emergentist Worldview," in *From Complexity to Life: On the Emergence of Life and Meaning*, ed. Niels Henrik Gregersen (Oxford: Oxford University Press, 2003), 15.

5. The incongruity between more progressive forms of emergence and the standard neo-Darwinian views such as Stephen Jay Gould's that focus on contingency is captured by Paul Davies in his claim that "[i]t is the essence of Darwinism that nature has no foresight—it cannot look ahead and anticipate the adaptive value of specific complex features." Paul Davies, "Complexity and the Arrow of Time," in Gregersen, *From Complexity to Life*, 88. For Davies, progressive trends sound "unnervingly teleological," though astrophysics lends some support to such an assumption.

6. Stephen J. Gould, *Time's Arrow: Time's Cycle-Myth and Metaphor in the Discovery of Geological Time* (Cambridge, MA: Harvard University Press, 1987); James Hutton, "Theory of the Earth," *Transactions of the Royal Society of Edinburgh* 1 (1788): 209–305. I am indebted for this insight to geologist Professor Alan Werritty, University of Dundee, who provided notes on James Hutton entitled "Deep Time and Sacred Space" for a guided tour of Arthur's Seat, Edinburgh, at the conference "Faith Communities and Environmental Activism," Edinburgh University, May 18th, 2017.

7. Simon Conway Morris, *Life's Solution: Inevitable Humans in a Lonely Universe* (Cambridge University Press, 2006).

8. Davies, "Complexity," 88.

9. Conway Morris, *Life's Solution*.

10. Terrence Deacon, *The Symbolic Species: The Co-Evolution of Language and the Brain* (New York: W. W. Norton, 1997); Ian Tattersall, "An Evolutionary Framework for the Acquisition of Symbolic Cognition by Homo sapiens," *Comparative Cognition & Behavior Reviews* 3 (2008): 99–114.

11. Philip G. Chase and Harold L. Dibble, "Middle Paleolithic Symbolism: A Review of Current Evidence and Interpretations," *Journal of Anthropological Archeology* 6 (1987): 263–96.

12. Robert H. Gargett, "Grave Shortcomings: The Evidence for Neanderthal Burial," *Current Anthropology* 30.2 (1989): 157–90.

13. Julian Riel-Salvatore and Geoffrey Clark, for example, tried to see if there was a connection between burial practice and cognitive abilities. There seemed to be no such correlation present, leaving open the criticism that the results so far had failed to indicate any definite intentional burial activity. Julian Riel-Salvatore and Geoffrey A. Clark, "Grave Markers: Middle and Early Upper Paleolithic Burials and the Use of Chronotypology in Contemporary Paleolithic Research," *Current Anthropology* 42.4 (2001): 449–79.

14. Sarah Tarlow, "Emotion in Archaeology," *Current Anthropology* 41.5 (2000): 713–46.

15. Marc Kissel and Agustín Fuentes, "From Hominid to Human: The Role of Human Wisdom and Distinctiveness in the Evolution of Modern Humans," *Philosophy, Theology and the Sciences* 3.2 (2016): 217–44.

16. Kissel and Fuentes, "From Hominid to Human"; Marc Kissel and Agustín Fuentes, "Semiosis in the Pleistocene," *Cambridge Archeological Journal* 3.21 (2017): 1–16.

17. Curtis W. Marean, "An Evolutionary Anthropological Perspective on Modern Human Origins," *Annual Review of Anthropology* 44 (2015): 533–56.

18. Agustín Fuentes, "Integrative Anthropology and the Human Niche: Toward a Contemporary Approach to Human Evolution," *American Anthropologist* 117 (2015): 302–15; Agustín Fuentes, "The Extended Evolutionary Synthesis, Ethnography and the Human Niche: Towards an Integrated Anthropology," *Current Anthropology* 57 Reintegrating Anthropology From Inside Out: Wenner-Gren Symposium Supplement (2016): S13–S26.

19. Alan Mittleman, *Human Nature and Jewish Thought* (Princeton: Princeton University Press, 2015).

20. Mittleman, *Human Nature*, 14–15. Of course, according to Hans Jonas, the basic tenets of evolutionary theory as expressed in Darwinian views of evolution by natural selection are just as dogmatic in philosophical terms; the difference is that, in Jonas' reading, neo-Platonic views are replaced by a new emphasis on contingency, so the ultimate driver of evolution by natural selection appears to the outsider to be that of random mutations, even if external conditions act as constraining "necessity." Gone is any sense of what he calls the phenomenon of life itself. Eventually, such views narrowed further to contemporary existentialist positions (Hans Jonas, *The Phenomenon of Life* [Evanston, IL: Northwestern University Press, 1966]). An evolutionary biologist reading this account would find his views correct in basic outline, but bald in detail. For example, teleology has not disappeared quite to the extent that Jonas suggests

here, and many biologists are just as ready to admit inadequacies of Darwinian theory as metaphysics, except, of course, the ultra-Darwinians like Richard Dawkins or Daniel Dennett. Emergence and top down causation is a counter-stream to the kind of philosophical portrait of Darwinian biology that he presents, though the progressive element in emergence can be just as troubling as the inadequacy stemming from too strong a focus on contingency, as noted earlier.

21. Mittleman, *Human Nature*, 14.
22. Ibid., 37.
23. Ibid., 47.
24. Ibid., 48.
25. Ibid., 49.
26. Ibid.
27. Ibid., 52.
28. Ibid., 53.
29. Ibid., 57.
30. Ibid., 58.
31. Ibid., 60.
32. Thomas Aquinas, *Summa Theologiae*, ed. T. Gilby and T. C. O'Brien, 60 vols. (London: New Blackfriars, 1964–1973), 2a2ae 49.
33. Josef Pieper, *The Four Cardinal Virtues* (Notre Dame: University of Notre Dame Press, 1966), 26.
34. Donna Hart and Robert W. Sussman, *Man the Hunted: Primates, Predators, and Human Evolution*, 2nd ed. (Boulder, CO: Westview Press, 2009); Robert Sussman, "Why the Legend of the Killer Ape Never Dies: The Enduring Power of Cultural Beliefs to Distort Our View of Human Evolution," in *War, Peace, and Human Nature: The Convergence of Evolutionary and Cultural Views*, ed. Douglas Fry (New York: Oxford University Press, 2013), 97–111.
35. Aquinas, *Summa Theologiae*, 2a2ae 47.9.
36. Aquinas, *Summa Theologiae*, 2a2ae 49.6.
37. Ibid., 2a2ae 47.1.
38. Ibid., 2a2ae 49.6.
39. Ibid.
40. Ibid., 2a2ae 55.8.
41. Martin Seligman, "Preface," in *Homo Prospectus*, ed. Martin Seligman, Peter Railton, Roy F. Baumeister, and Chandra Sripada (Oxford: Oxford University Press, 2016), ix.
42. Peter Railton, "Introduction," in *Homo Prospectus*, ed. Seligman et al., 7.
43. Mary Carruthers, *The Book of Memory: A Study of Memory in Medieval Culture* (Cambridge: Cambridge University Press, 1990), 193.
44. Railton, "Introduction," 5.
45. Aquinas, *Summa Theologiae*, 2a2ae 47.1.
46. See Peter Railton, "Intuitive Guidance: Emotion, Information and Experience," in *Homo Prospectus*, ed. Seligman et al., 33–85; Chandra Sripada, "Deliberative Guidance: Intuitive Guidance in the Counterfactual Mode," in *Homo Prospectus*, ed. Seligman et al., 87–101.
47. Aquinas, *Summa Theologiae*, 2a2ae 47.1.
48. Ibid., 2a2ae 48.
49. See Aquinas, *Summa Theologiae*, 2a2ae 49.2 for insight/intelligence and 2a2ae 49.5 for reasoned judgment.
50. Ibid., 2a2ae 49.2.
51. Ibid.
52. Ibid.
53. Ibid., 2a2ae 49.5. For further discussion of the distinction between angels and human mortals in Thomas, and its value for considering the specificity of human wisdom, see Adam Willows' chapter in this volume.
54. Railton, "Intuitive Guidance," 73.

55. Aquinas, *Summa Theologiae,* 2a2ae 47.1.

56. According to this view, morality and religion were separate, the first religions portraying immoral gods, with monotheistic Gods appearing much later as a result of the need to trust in expanded networks of cooperation. Roy Baumeister, "Collective Prospection: The Social Construction of the Future," in *Homo Prospectus*, ed. Seligman et al., 151. See also Ara Norenzayan, *Big Gods: How Religion Transformed Cooperation and Conflict* (Princeton: Princeton University Press, 2013). For an excellent collection of review articles see *Religion, Brain & Behavior* 5.4 (2015). The review by Agustín Fuentes is particularly pertinent in that he recognizes that Ara Norenzayan presupposes that interventionist and moralizing perceptions of God correlate with increase in population size and hyper complex social structures. Unlike other proposals, Norenzayan presupposes that Big Gods arose through cultural evolution rather than evolution of specific cognitive capacities. Yet Norenzayan's overall thesis still relies on finding an explanation for why cooperative groups could emerge based on the premise that kin selection and reciprocal altruism are the dominant drivers of complex cooperative sociality. Fuentes, in a critical move, rejects that model of evolution, and proposes instead that basal sociality is not something that requires explanation since it was integral to the evolutionary processes from the start. There is therefore no need to "invent" a Big Gods hypothesis to explain the flaws in the ability of standard Neo-Darwinian explanations of hyper-cooperation. He also recognizes that the experience of those who are in religious communities is not as simple as the Big Gods thesis proposes. Agustín Fuentes, "Hypercooperation Is Deep in Our Evolutionary History and Individual Perception of Belief Matters," *Religion, Brain & Behavior* 5.4 (2014): 266–342. From a theological perspective the idea that God *both* emerges from human experience *and* is also a construct that enables control of such groups in an evolutionarily important way makes little sense. Rather, the cognitive capacity for conscious belief in God may evolve, but the receptive experience of God is primary and one that promotes rather than inhibits human flourishing and affectivity.

57. Aquinas, *Summa Theologiae*, 1a 22.4.

58. Joseph Soloveitchik, *The Emergence of Ethical Man*, ed. Michael Berger (Jersey City: Ktav Publishing House, 2005). Published posthumously.

59. Ibid., 78.

60. Ibid., 78–79.

61. Ibid., 79.

62. Aquinas, *Summa Theologiae,* 2a2ae 49.1.

63. Ibid.

64. Ibid., 1a2ae 56.5.

65. Ibid., 1a 78.4.

66. Ibid.

67. Ibid.

68. Mary Carruthers and Jan M. Ziolkowski, "General Introduction," in *The Medieval Craft of Memory: An Anthology of Texts and Pictures* (Philadelphia: University of Pennsylvania Press, 2002), 3.

69. Thomas Aquinas wrote a detailed commentary on Aristotle's *De memoria et reminiscentia* (On memory and recollection) which brings out this point forcefully. Aristotle's text was part of the *Parva naturalia* (little books on natural science), one of nine topics covered that were intended to complement his well-known larger text *De anima* (On the soul). The *Parva naturalia* was itself subdivided into *parvi libri* and *parvi libri naturalis*. For discussion see the introduction by Jan M. Ziolkowski to Thomas Aquinas, "*Commentary* on Aristotle, *On Memory and Recollection*," trans. John Burchill, in *The Medieval Craft of Memory*, ed. Carruthers and Ziolkowski, 153–56.

70. The individual characteristics flow more readily once language arises, which of course the medievalists assumed. White, for example, could be associated with "snow," "milk," or "lilies." Without language it is hard to appreciate how an abstract universal concept like "white" could be stored in the memory. See Carruthers, *Book of Memory*, 64 for a discussion of medieval techniques in associative recollection.

71. Carruthers and Ziolkowski, "General Introduction," 2.
72. Thomas Aquinas, "Lesson 1, §298," in "*Commentary* on Aristotle," 156.
73. Aquinas, "Lesson 1, §299," in "*Commentary* on Aristotle," 156.
74. Aquinas, "Lesson 8, §399," in "*Commentary* on Aristotle," 186.
75. Aquinas, "Lesson 8, §402, 408," in "*Commentary* on Aristotle," 187–88.
76. Hans Jonas, "Image Making and the Freedom of Man," in *The Phenomenon of Life* (Evanston, IL: Northwestern University Press, 1966), 157–75.
77. Ibid., 158.
78. Ibid., 166.
79. Ibid., 170.
80. Ibid.
81. Ibid., 171.
82. Ibid.
83. Ibid., 172.
84. Ibid.
85. Michael Tomasello, *A Natural History of Human Morality* (Cambridge, MA: Harvard University Press, 2016), 141.
86. I discuss this further in Celia Deane-Drummond, "Empathy and the Evolution of Compassion: From Deep History to Infused Virtue," *Zygon: Journal of Religion and Science* 52.1 (2017): 258–78.
87. Tomasello argues that interdependence is primary and that this is followed by cognitive adaptations in the form of shared intentionality. He also speculates that second personal agency between two persons based on mutual respect and what is considered fair, a characteristic he terms "deservingness," was more likely to have arisen prior to joint commitment, where persons regulate their behavior on the basis of a cooperative sense of obligation and rationality. Tomasello, *A Natural History*, 1523.
88. Marc Kissel has collated a fascinating database of evidence in the archeological literature for complex tool making and artifacts that have non-utilitarian use, such as bone technology, engravings on shells, red ocher, production of large tools known as blades, beads, and so on, in very early humans. The astonishing result is that there are preliminary signs of such activity very deep indeed, as far back as *Homo erectus*, and every "marker" appears at least 300,000 years ago, long before anatomically modern humans first appeared around 200,000 years ago. See Kissel and Fuentes, "From Hominid to Human."

BIBLIOGRAPHY

Aquinas, Thomas. "*Commentary* on Aristotle, *On Memory and Recollection*." Translated by John Burchill. In *The Medieval Craft of Memory*, edited by Mary Carruthers and Jan Ziolkowski, 153–87. Philadelphia: University of Pennsylvania Press, 2002.

———. *Summa Theologiae*. Edited by T. Gilby and T. C. O'Brien. 60 vols. London: New Blackfriars, 1964–1973.

Barbour, Ian. "Neuroscience, Artificial Intelligence and Human Nature: Theological and Philosophical Reflections." In *Neuroscience and the Person: Scientific Perspectives on Divine Action*, edited by Robert J. Russell, Nancey Murphy, Theo C. Meyering, and Michael A. Arbib, 249–80. Notre Dame: University of Notre Dame Press, 1999.

Baumeister, Roy. "Collective Prospection: The Social Construction of the Future." In *Homo Prospectus*, edited by Martin Seligman, Peter Railton, Roy F. Baumeister, and Chandra Sripada, 133–56. Oxford: Oxford University Press, 2016.

Carruthers, Mary, and Jan M. Ziolkowski. "General Introduction." In *The Medieval Craft of Memory: An Anthology of Texts and Pictures*, edited by Mary Carruthers and Jan Ziolkowski, 1-31. Philadelphia: University of Pennsylvania Press, 2002.

———, eds. *The Medieval Craft of Memory: An Anthology of Texts and Pictures*. Philadelphia: University of Pennsylvania Press, 2002.

Carruthers, Mary. *The Book of Memory: A Study of Memory in Medieval Culture*. Cambridge: Cambridge University Press, 1990.

Chase, Philip G., and Harold L. Dibble. "Middle Paleolithic Symbolism: A Review of Current Evidence and Interpretations." *Journal of Anthropological Archeology* 6 (1987): 263–96.

Clayton, Philip. *God and Contemporary Science*. Edinburgh: Edinburgh University Press, 1997.

Conway Morris, Simon. *Life's Solution: Inevitable Humans in a Lonely Universe*. Cambridge: Cambridge University Press, 2006.

Davies, Paul. "Complexity and the Arrow of Time." In *From Complexity to Life: On the Emergence of Life and Meaning*, edited by Niels Henrik Gregersen, 72–92. Oxford: Oxford University Press, 2003.

———. "Introduction: Toward an Emergentist Worldview." In *From Complexity to Life: On the Emergence of Life and Meaning*, edited by Niels Henrik Gregersen, 3–18. Oxford: Oxford University Press, 2003.

Deacon, Terrence. *The Symbolic Species: The Co-Evolution of Language and the Brain*. New York: W. W. Norton, 1997.

Deane-Drummond, Celia. "Empathy and the Evolution of Compassion: From Deep History to Infused Virtue." *Zygon: Journal of Religion and Science* 52.1 (2017): 258–78.

Fuentes, Agustín. "The Extended Evolutionary Synthesis, Ethnography, and the Human Niche: Towards an Integrated Anthropology." *Current Anthropology* 57 Reintegrating Anthropology From Inside Out: Wenner-Gren Symposium Supplement (2016): S13–S26.

———. "Hypercooperation Is Deep in Our Evolutionary History and Individual Perception of Belief Matters." *Religion, Brain & Behavior* 5.4 (2014): 266–342.

———. "Integrative Anthropology and the Human Niche: Toward a Contemporary Approach to Human Evolution." *American Anthropologist* 117 (2015): 302–15.

Gargett, Robert H. "Grave Shortcomings: The Evidence for Neanderthal Burial." *Current Anthropology* 30.2 (1989): 157–90.

Gould, Stephen J. *Time's Arrow: Time's Cycle-Myth and Metaphor in the Discovery of Geological Time*. Cambridge, MA: Harvard University Press, 1987.

Gregersen, Niels Henrik, ed. *From Complexity to Life: On the Emergence of Life and Meaning*. Oxford: Oxford University Press, 2003.

Hart, Donna, and Robert W. Sussman. *Man the Hunted: Primates, Predators, and Human Evolution*, 2nd edition. Boulder, CO: Westview Press, 2009.

Hutton, James. "Theory of the Earth." *Transactions of the Royal Society of Edinburgh* 1 (1788): 209–305.

Jonas, Hans. *The Phenomenon of Life*. Evanston, IL: Northwestern University Press, 1966.

Kissel, Marc, and Agustín Fuentes. "From Hominid to Human: The Role of Human Wisdom and Distinctiveness in the Evolution of Modern Humans." *Philosophy, Theology and the Sciences* 3.2 (2016): 217–44.

———. "Semiosis in the Pleistocene." *Cambridge Archeological Journal* 3.21 (2017): 1–16.

Marean, Curtis W. "An Evolutionary Anthropological Perspective on Modern Human Origins." *Annual Review of Anthropology* 44 (2015): 533–56.

Mittleman, Alan. *Human Nature and Jewish Thought*. Princeton: Princeton University Press, 2015.

Murphy, Nancey. "Introduction." In *Evolution and Emergence: Systems, Organisms, Persons*, edited by Nancey Murphy and William R. Stoeger, 1–18. Oxford: Oxford University Press, 2007.

Norenzayan, Ara. *Big Gods: How Religion Transformed Cooperation and Conflict*. Princeton: Princeton University Press, 2013.

Pieper, Josef. *The Four Cardinal Virtues*. Notre Dame: University of Notre Dame Press, 1966.

Railton, Peter. "Introduction." In *Homo Prospectus*, edited by Martin Seligman, Peter Railton, Roy F. Baumeister, and Chandra Sripada, 3–32. Oxford: Oxford University Press, 2016.

———. "Intuitive Guidance: Emotion, Information and Experience." In *Homo Prospectus*, edited by Martin Seligman, Peter Railton, Roy F. Baumeister, and Chandra Sripada, 33–85. Oxford: Oxford University Press, 2016.

Riel-Salvatore, Julian, and Geoffrey A. Clark. "Grave Markers: Middle and Early Upper Paleolithic Burials and the Use of Chronotypology in Contemporary Paleolithic Research." *Current Anthropology* 42.4 (2001): 449–79.

Seligman, Martin. "Preface," In *Homo Prospectus*, In *Homo Prospectus*, edited by Martin Seligman, Peter Railton, Roy F. Baumeister, and Chandra Sripada, ix-xiv. Oxford: Oxford University Press, 2016.

Seligman, Martin, Peter Railton, Roy F. Baumeister, and Chandra Sripada, eds. *Homo Prospectus*. Oxford: Oxford University Press, 2016.

Soloveitchik, Joseph. *The Emergence of Ethical Man*, edited by Michael Berger. Jersey City: Ktav Publishing House, 2005.

Sripada, Chandra. "Deliberative Guidance: Intuitive Guidance in the Counterfactual Mode." In *Homo Prospectus*, edited by Martin Seligman, Peter Railton, Roy F. Baumeister, and Chandra Sripada, 87–101. Oxford: Oxford University Press, 2016.

Sussman, Robert. "Why the Legend of the Killer Ape Never Dies: The Enduring Power of Cultural Beliefs to Distort Our View of Human Evolution." In *War, Peace, and Human Nature: The Convergence of Evolutionary and Cultural Views*, edited by Douglas Fry, 97–111. New York: Oxford University Press, 2013.

Tarlow, Sarah. "Emotion in Archaeology." *Current Anthropology* 41.5 (2000): 713–46.

Tattersall, Ian. "An Evolutionary Framework for the Acquisition of Symbolic Cognition by Homo sapiens." *Comparative Cognition & Behavior Reviews* 3 (2008): 99–114.

Tomasello, Michael. *A Natural History of Human Morality*. Cambridge, MA: Harvard University Press, 2016.

Epilogue

Questions and Puzzles in Evolutionary Anthropology

TEN

Manipulating Materials, Bodies, and Signs

How the Ecology of Creative Problem Solving, Tool Manufacture, and Imaginative Sociality Set the Context for Language in the Later Pleistocene Human Niche

Agustín Fuentes

Language is critical to the contemporary human niche. Key aspects of human evolution over the past fifteen to twenty millennia are tied to linguistic capacities and manipulation (e.g., gender, political, and economic hierarchies, expansive kinship structures, group identities, broad-scale domestication, inter-societal conflict and peace, etc.). But the evolution of language, its first appearances, and anything that we might term "proto-language" are areas where a hypothesis-driven scientific analysis is difficult. Early linguistic behavior is extremely difficult to detect in the archeological record or fossil data, and genetic and physiological assessments (contemporary and ancient) rarely provide specific insights that prove explanatory for the emergence of language. Here I (following many others) suggest that there are other options for understanding the emergence of language per se in Pleistocene *Homo*. Rather than attempting to pinpoint or explain the emergence of language itself, I propose that an approach examining the development of a human wisdom (which I operationally define as the capacity to create and manipulate material items and social relations with increasing skill and imagination) might inform our models of the human niche in a manner that facilitates a better understanding of the contexts in which language emerged. This essay does not attempt to describe the evolution of language, or proto-language; rather, it looks to describe and interpret aspects of the human niche in the later Pleistocene that provided the social, ecological, and neurobiological contexts within which language could develop. In short, the emergence of a human wisdom facilitated the eventual emergence of human language.

LANGUAGE, COMMUNICATION, AND WISDOM

Put simply, human pragmatic communication is an exaptation of—and hence is evolutionarily continuous with—human social intelligence.[1]

The phylogenetic link between human language and other forms of primate communication is not in the vocal or gestural communicative practices but in the deep and rich patterns of social cognition. Human wisdom, as defined above, precedes language in the human niche. Human language, per se, emerges from a multiplicity of complex feedback system, signs, mental/material manipulations, and behavioral ecologies in the context of a rich material and perceptual culture. Obviously, many components that enable language have deep roots in the primate lineage (social complexity, face-to-face visually emphasized communication, large and dynamic social groups, a high degree of coordinated activities, etc.), but language itself is a peculiarly human combination of primate legacies, hominin novelties, and human distinctions.

Here a note regarding the two main communication models (*code* and *ostensive-inferential*) is worthwhile. I draw on Thomas Scott-Phillips' overview for a concise summary:

> Code model: "Information is encoded, sent along a communication channel, and then decoded at the other end. If the algorithms for encoding and decoding are appropriately calibrated to one another, then what is encoded at one end is identical to what is decoded at the other end. The result is that information has been transferred from a signaler to a receiver. This way of thinking about communication is called the code model."[2]

> Ostensive-inferential model: "The production and comprehension of signals does not involve encoding and decoding of a message but instead the provision and interpretation of evidence of intentions. The receiver (target) must then take this evidence and draw an inference about the signaler's intended meaning. Signalers and receivers thus perform distinct but complementary jobs in ostensive-inferential communication: the former provide evidence (this is ostension); the latter interpret it (this is inference)."[3]

One can envision such things as monkey alarm calls for predators or specific "danger" messages carried by bright colors and markings of skunks or snakes as examples of the code model. However, human language, with its complex symbolic, linguistically, and semantically diverse modalities of message construction and transfer, is an intensive ostensive-inferential system. Human language is based on a complex symbolic infrastructure and reliant (at least today) on both vocal and visual elements (speech and writing). Thus, understanding the emergence of language capacities and language itself is best undertaken via understanding the processes and patterns of emergence of human social cognition, shared intentionality, teaching, and symbolic signs, all of which are necessary infrastructure from which the current intensive ostensive-inferential system we call "language" could develop.

OTHER PRIMATES

Does looking to the other primates assist in this endeavor? Only partially. Cognitively anthropoid primates likely have a mirror system[4] matching observation and execution of hand actions. They have a simple imitation system for manual actions which is tied to a system of novel communicative gestures.[5] In general, primates have complex code-based communication (gestural) systems and a relatively limited vocal communication repertoire. Primate vocal communication takes the form of vocalizations that act to convey prosocial or agonistic signals, such as reassurance or threats to other group members. Most species also have warning/alarm calls to alert other group members to threats. In some cases these warning calls are divided into discrete categories for different types of threats as in the classic examples of vervet monkey alarm calls for terrestrial, aerial, and serpent predators.[6] Other primates have long calls or related forms of long distance vocal communication, like that of howler monkeys. A few species of primates, such as the gibbons (smaller apes), and possibly indriids (a type of lemur), are distinctive in having broader vocal communicative repertoires. Their calls are structured and melodic and have a range of specific content elements reflecting a more structured suite of information than most primate vocalization. However, no primates approach the depth and density of information transfer that characterizes human language. In fact, vocal communication is not the main mode of information transfer in the majority of primates; most species of anthropoid (monkeys and apes) primate (and many other socially complex mammals) have the core of their behavioral communication content tied to behavioral actions relating to negotiating social complexity and social relations. Moreover, nearly all of it falls into the code model.

It is argued that great apes (chimpanzees, gorillas, and orangutans) do express a range of examples of ostensive communication (intentionally overt communication), and do so more than other primates, but they do not have language.[7] Interestingly, those apes cross-fostered by humans, inculcated in the human niche, and taught to interact using human-designed communicative systems, display more types of, and/or larger capacities for, ostensive communication.[8] To me this suggests a critical role for the human niche in providing the ecological, social, and structured context and stimulus for complex metapsychological behaviors (theory of mind and expanded shared intentionality) that are required for the emergence and use of language.[9]

PLEISTOCENE *HOMO*: INITIATION OF THE HUMAN NICHE

We know the hominin baseline for social complexity. This baseline is the set of behavioral patterns and/or adaptations that characterize the lineages from which the genus *Homo* arose. Such a baseline can be constructed by examining all of the living ape species and *Homo sapiens*, and noting what all groups share in common. On this basis, the patterns which we can comfortably assume that the LCA (the last common ancestor between Hominins and other Hominines [African apes]) exhibited were some form of community (fission-fission multi-male and multi-female) social organization, increased social cognition relative to other primates, frequent social reciprocity, local social tradition and innovation ("culture"), and simple stone tool use.[10] We might also assume—given the patterns observed in the remaining great apes, modern humans, and the hominin lineage—a strong role for social niche construction.[11]

Over the course of our Pleistocene evolution, a range of distinctive cognitive and behavioral capacities developed. These include expanded cultural innovation via manipulation of extra-somatic materials (tools and others) and accumulation of cultural complexity via an autocatalytic process involving feedback between creativity and transmission. These features result in a distinctive human socio-cognitive niche,[12] a co-evolutionary interdependence between the ecological, cognitive, and neural systems, and skill transmission via an apprentice model.[13] This dynamic provides the context for the development of a "language-ready brain"[14] and lays the template for an increasingly complex system of ostensive-inferential communication.[15]

Niche construction and feedback/feed-forward loops involving neurobiology, ecology, and behavior are central to the evolution of tool use and creation in primates and possibly other organisms. The role of tool creation/manipulation and its impact is ratcheted up significantly in *Homo*. Tool creation and use had substantive influence on human morphology, neuroanatomy, ecology and developmental processes.[16] This makes it especially interesting in understanding the context for the development of the infrastructure necessary for language.

Humans, unlike other tool using animals like the great apes, have rapid cultural innovation and accumulation via an autocatalytic process that involves feedback between innovation, use, and mode of transmission.[17] This is a dynamic system wherein innovations (e.g., ideas and behaviors) and elements (e.g., tool types, food processing patterns) are developed, acquired, and lost depending on social densities, transmission fidelity, and social complexity. In these contexts, feedback processes within the social niche are a core mechanism for innovation, transition, and modification of materials, behaviors, and cognitive capabilities.[18] In the material context (e.g., stone tools), novel innovation is often generated by combining existing technologies and their constituent elements with high-density information transfer and complex social cognitive capacities. Take the emergence of the earliest Acheulean tools around 1.6 million years ago. A key pattern underlying these innovations is greater pre-modification of the core (the rock from which a flake that is going to be made into a final tool is knocked off) to create specific patterns in the surface of the core such that the likelihood of getting a really good flake is improved—this requires more pre-planning and a more multi-layered visualization of what the end result could be. This is different from the earlier Oldowan styles (dating to ~2.4–1.8 million years ago) in that the core is modified specifically in order to enable better flake detachments. There are even more steps of preparation, of envisioning multiple stages to the process, before it is even begun. How did such dense, technical information get transferred and shared?

Social and material innovation can emerge individually or via cooperation, but intensive social coordination and shared intentionality can combine to develop high accuracy of transmission. In such scenarios the density and resilience of social connections can act to increase the resilience of innovations.[19] For example, Middle Pleistocene stone tool technologies diversified substantially between 500,000 and 200,000 years ago when cognitive and cooperative patterns in multiple populations of the genus *Homo* were likely accompanied by slight increases in population densities and frequencies of intergroup social interactions.[20] It is possible that during this time period instances of innovation, and the chances of such innovations being maintained and shared, increased (due to more interactions between more groups/individuals), while the likelihood of new innovation extinction decreased, as novel patterns and processes were more and more widely shared.[21]

Emerging from the increasingly dense conception and transmission of innovation alongside substantive neurological and behavioral plasticity is the human socio-cognitive niche,[22] a niche wherein the "language-ready brain"[23] is able to emerge. This niche is a cognitive and behavioral configuration that is derived relative to those of previous hominins as it fosters a central role for increasingly complex communicative interfaces involving dense information exchange via use of signs, vocal and gestural behaviors, and the forging of diverse social bonds. These communicative processes become a core part of how humans deal with ecological and social challenges. This approach is complementary with the social brain and extended cognition hypotheses[24] and the broader niche construction approaches to human evolution.[25] However, it has not often been connected directly to the emergence of language capacities.[26]

How does this niche, across the Pleistocene, create the template for such increasingly complex communicative systems?

The human socio-cognitive niche is a dynamic interdependence between ecological, cognitive, and neural processes and represents a context for the reshaping of neural architecture over human evolutionary history.[27] Pleistocene humans began using "motor tools," such as stone and wood, to extend somatic motor capabilities expanding their capacities to acquire and process animals and plants. There is a marked increase in dietary quality and breadth and a lowering of processing costs in the latter half of the Pleistocene for the genus *Homo*. Today, and likely in the past, humans use "sensory tools" to extend or externalize our existing sensory organs, enhancing our sight or hearing beyond the immediate somatic potential of our eyes and ears. We also use "symbolic tools" in much the same way: externalized and shared memories, symbolic meanings placed on the landscape, and even material items with social meaning can all act to unite group members. The timing and patterns of the emergence of such processes is not fully clear from the archeological record; however, recent overviews demonstrate that archeological evidence of such a process is available in the Pleistocene much earlier than the appearance of overt symbolic elements generally tied to linguistic capacities (e.g., cave art ~40,000 ya).[28] The capacity to increasingly use more and more diverse forms of "tools" (motor, sensory, and symbolic) offers a rich array of options for incorporation of diverse elements into the developing communication system in *Homo*, a system that generates both somatic and extra-somatic complexity as it expands via iterative feedback processes.

The increases in brain size and complexity, the extended childhood period, and the plasticity of brain development over the trajectory of the Pleistocene likely interfaced, in multiple feedback loops, with the innovation and increased use of extrasomatic materials and behavioral options in the behavioral ecology of *Homo*. The use of tools, the control of fire, and the creation of material symbols involve a dynamic feedback between action, perception, and neural structures, expanding the potential for information acquisition and problem solving. This, in turn, facilitates the potential for rapid and dramatic changes in patterns of interactions with local ecologies and the development of neural architecture.[29]

For example, cultural innovation and accumulation, the socio-cognitive niche, and the interaction between "tool" use and neurological systems involve high intensity and broad bandwidth of information transfer.[30] From at least the Acheulean phase II forward (~500,000 ya), acquisition and reliable replication of tool making skills has required the following:

a. observation of skilled tool makers
b. access to full and partial templates of the tools themselves
c. some form of dense information transfer: tool making skills are acquired by task decomposition and the ordering of skill acquisition.[31]

Sterelny proposed *the apprentice model* as a central feature of the human niche from at least the middle Pleistocene onward.[32] The apprentice model is a hybrid learning system in which central social and material skills are acquired by combining information from the social world and the physical-biological world. Individuals learn by "doing" in an environment seeded with informational resources but not necessarily explicit instruction or formalized institutions. Incremental construction, emulation, and passive and active information transfer in the context of a highly social cooperative group (with cognitive complexity and shared intentionality) can generate high fidelity replication and broad bandwidth information transfer. There is substantive archeological evidence that such outcomes were occurring by 3–500,000 years ago (in stone tool creation/use, fire use, and the emergence of complex hunting).

By the terminal Pleistocene (the last ~100–200,000 years) the fossil and material records demonstrate that humans have a distinctive socio-cognitive niche and a cumulative and ratcheting culture impacting neural architecture, and that a dense, symbolic communication system is in place. But is there language?

Michael Arbib argues that, by the time our species (*Homo sapiens*) appeared, the human niche contained a complex imitation system and was capable of developing novel communicative gestures in combination with increasingly dense cognitive capacities and the ability to use iconic and indexical signs (and other signs types within the Peircean context—see Kissel in this volume).[33] This facilitated the emergence of a set of sign semantics which in turn facilitated an evolutionary path (behavioral-ecological-neurobiological) for "new control patterns for the vocal articulators that made possible the incorporation of non-innate vocal gestures into the communicative repertoire, yielding a capacity for proto-speech."[34] In the last few hundred thousand years of human evolution, groups existing in this increasingly information- and materially-dense feedback process could develop novel manual and vocal gestures in order to convey an expanding set of meanings, meanings infused with symbolic information, and represented increasingly by organized vocal and gestural communicative actions.[35]

A NOTE ON MOVEMENT, RHYTHM, AND STORYTELLING

There are other avenues proposed for the emergence of increasingly complex communicative systems in humans, systems not exclusively focused on sign, meaning, and complex motor-skill-neurobiology feedback systems. These ideas relate to language but offer us a few other avenues for its development.

The archeologist Stephen Mithen argues that musicality is a core part of being human and that it might have played a central role in the emergence of language. Mithen proposes that it was the ability to create melody with voices and via the use of material items that acted as the precursor to spoken language.[36] In this model, the use of melodic sequences is deployed to convey information and tell stories, and acts to offer cohesion for groups of middle Pleistocene members of the genus *Homo*. These melodic patterns acted as a form of protolanguage of musical and emotional expression. Mithen suggests that Pleistocene *Homo* developed a system of communication he calls "Hmmmm."[37] The system was a mixture of gesture and melody, and was the precursor to language. Mithen envisions that these members of the genus *Homo* made extensive use of their physiological range of pitch, rhythm, and even the development of melody to communicate information, express emotion, and elicit emotional responses in other individuals.

Mithen notes that other primates use vocalizations, gestures, and body postures for communication and thus, logically, suggests that our ancestors shared those deep primate capacities. As noted earlier, many non-human primates use vocalizations, even melodic ones, to communicate with one another about emotional states and alarms/threats, and in the creation and maintenance of social bonds. All the apes use such vocal behavior and gibbons and siamangs use highly structured, melodic calls (often referred to as "songs") as their main mode of communication. However, Mithen goes further than simple phylogenetic correlations and asserts that there may be an inherent musicality to primate vocalizations, which in turn formed the basis for hominins who used such a capacity even before the appearance of our own genus *Homo*. We have established that, as the members of the genus *Homo* developed more complex cognitive and behavioral patterns over the first 1.5 million years or so of their evolution, they appear to have developed more complex modes of communication, as evidenced by the appearance and spread of increasingly complex tools, power scavenging, and cooperative parenting, and some forms of teaching. Such advances, argues Mithen, would not have been possible without a ratcheting system of communication. Mithen argues that part of this new system of communication was the increased use of "sound synesthesia," in which vocalizations are used to represent the size of things and/or their movement—for example, the whistling sound humans often make when describing something falling to the ground (pheeeeeeeeooowww, boom!) or the sounds we use to describe the galloping of a horse (gaalump, gaalump, gaalump). In this scenario of the emergence of more complex types of material expression in *Homo* groups, the ratcheting up of hunting and fire, and the increasingly complex social lives in the direct ancestors of modern humans, this protolanguage system becomes progressively modified into more complex melodic aspects (music and song) and more particulate vocal aspects (more sounds that had discrete meanings). Such changes, Mithen suggests, lead eventually to the emergence of language. For Mithen, music and language go hand in hand, and he suggests that there is good neurobiological evidence that the two are tightly interconnected.

A related argument is put forward by the philosopher Maxine Sheets-Johnstone.[38] She argues that dance and movement are central to human cognitive development and the ways in which humans have created complex emotional behavioral and communication systems. Sheets-Johnstone places an emphasis on dance and the use and coordination of movement to convey empathy, meaning, and content. She makes a similar argument to Mithen's in regard to melody and music. These arguments are also similar to those of the philosopher Merlin Donald, who argues for mimesis as central to the human experience and language,[39] mimesis being where humans began to rehearse and refine movements in controlled and systematic ways, which enabled them to reproduce the movements on command and imbue them with more and more complex meaning.[40]

This focus on melodic sounds and movement is enticing. However, the evidence in support of such positions comes mostly from modern human behavior, comparison with living primates and other animals, and a small amount of data from the archeological record. Evidence for music shows up late in the material record. The earliest clearly musical instruments date only to ~43,000 years ago and are bone flutes found at a few sites in the area of what is today Germany. It is only in the last 14,000 years or so that we start to find evidence of musical instruments more broadly, and paintings or art that clearly depict music or dance are also very recent. However, that does not mean that melodic sounds, especially percussion sounds ("drumming"), and dancing did not show up much earlier. Many human cultures use some form of drumming and some form of repetitive vocal chanting and melodic movement as part of their ways of expression. It is not a stretch to assume that early members of the genus *Homo* developed a pattern of using sound for increasingly complex communication early on. Chimpanzees are observed to drum on fallen trees to draw attention to themselves and many birds make highly sequential melodic pounding sounds in addition to their calls. It is not unlikely that earlier members of the genus *Homo* incorporated some sounds and the structures of those sounds into their increasingly complex lives and communication systems.

Another key factor in the emergence of language in humans is that of verbal and gestural storytelling. We have material evidence of this capacity in the art and paintings from approximately 40,000 years ago through to today, and possibly in the earlier appearances of symbolic materials,[41] but it remains a difficult aspect to pin down in relation to language.

For humans, a key aspect of language is the capacity for the conveying of substantial meaning and information via the use of metaphor, analogy, and dense description—what we call a story. This human capacity to gather a group together and—via sounds, gestures, and possibly visual aids—relate a set of ideas, events, hopes, and dreams is a key outcome of our evolution and the human socio-cognitive niche. Storytelling is how most humans navigate the world. On a daily basis, humans engage in this process to relate to others, to convey events, ideas, and experiences that are separate in time from the specific moment in which they are conveying them, and different in setting from the physical place in which they are relating them. But one could argue that whatever forms of pre-linguistic capacities our ancestors had they would have had to have been deployed and experimented with in order to further develop a more robust information exchange; this would be the emergence of the storytelling capacity. One could argue that this is among the key driving processes and patterns that accompany the emergence of human social cognition, shared intentionality, teaching and symbolic signs. Is the emergence of storytelling a form of distinctively human wisdom? If this "wisdom" is the capacity to create and manipulate material items and social relations with increasing skill and imagination, then yes, it might be.

The urge and drive to communicate increasingly complex and imaginative thoughts and the need to collaborate with others in intensive ways to make that happen is likely at the base of the evolution of what we term language today. But it is neither linear, nor only related to vocal communication, nor separable from the interweavings of the processes in the human socio-cognitive niche.

The emergence of language is the story of storytelling and the emergence of a kind of distinctively human way to describe, perceive, and modify the world. Nearly two million years ago, the earliest members of the genus *Homo* developed and shared the capacities to make Oldowan stone tools, and then coordinated power scavenging, the exploitation of new foods, and the development of new tool types. They moved on to hunting and fire and the use of ochre, the creation of engravings and figurines, the advent of domestication and agriculture, and the development of the visual arts. All the while our ancestors were moving, gesticulating, humming, grunting, dancing, singing, and eventually talking their way into the world we now occupy.

TENTATIVE CONCLUSIONS

> The implication of this for evolutionary issues is that because language is based on a foundation of ostension and inference and not code, then the appropriate comparisons to make are not with the natural codes of

great ape communication but rather with the social-cognitive mechanisms that underpin ostensive-inferential communication.[42]

I propose (as have many others) that we need to avoid the quest to find "proto-language" in other primates. Language is a distinctively human capacity and system and it is in the human niche developed over the Pleistocene that we need to seek its development. However, if what Scott-Phillips asserts is correct, then we can look to both the other primates (and other highly social mammals) and to the hominin fossil and archeological record to identify cues and patterns of socially driven, cognitively mediated communication—a particular intentional (ostentional) way of being in and with the world. Here the deployment of such a system of communication can be envisioned as a form of wisdom where information, social relations, shared intentionality, and cognitive pathways interact and co-construct one another. The roots of such a pattern of wisdom are not simply ingrained in the genes or wired into bodies as much as they are realized out of interactions with the phenomenal world and the processes of relations between organisms during the course of our lineage's evolution.[43]

It just may be the case that it was the interface between, on the one hand, our complex primate capacities and focus on social negation and the materials, bodies, and signs, creative problem solving, and, on the other, the imaginative sociality of the genus *Homo* in the Pleistocene, that set the context which made it possible for language to develop in the human niche.

NOTES

1. Thomas C. Scott-Phillips, "Nonhuman Primate Communication, Pragmatics, and the Origins of Language," *Current Anthropology* 56.1 (2015): 77.
2. Ibid., 58.
3. Ibid.
4. This is the neurobiological system in which actions by one individual, such as the movement of a limb, can stimulate neurological responses for that movement in another individual watching the first one, even if the watching individual does not move her limb. The neurological correlates of the motor function fire in response to seeing the motion in another. This is called mirroring.
5. Michael A. Arbib, "From Mirror Neurons to Complex Imitation in the Evolution of Language and Tool Use," *Annual Review of Anthropology* 40 (2011): 257–73.
6. R. M. Seyfarth, D. L. Cheney, and P. Marler, "Monkey Responses to Three Different Alarm Calls: Evidence of Predator Classification and Semantic Communication," *Science* 210 (1980): 801–3.
7. Scott-Phillips, "Nonhuman Primate Communication."
8. Michael Tomasello and Josep Call, "The Role of Humans in the Cognitive Development of Apes Revisited," *Animal Cognition* 7.4 (2004): 213–15. It is not known if these capacities are also found in those primates which are not raised by humans but which spend a good deal of time around them. However, anecdotally, this does not seem to be the case.

9. See, for example, Michael Tomasello, *Origins of Human Communication* (Cambridge, MA: MIT Press, 2008).

10. Agustín Fuentes, "Integrative Anthropology and the Human Niche: Toward a Contemporary Approach to Human Evolution," *American Anthropologist* 117.2 (2015): 302–15; Robert Foley and Clive Gamble, "The Ecology of Social Transitions in Human Evolution," *Philosophical Transactions of the Royal Society B* 364 (2009): 3267–79; N. M. Malone, A. Fuentes, and F. J. White, "Variation in the Social Systems of Extant Hominoids: Comparative Insight into the Social Behaviour of Early Hominins," *International Journal of Primatology* 33.6 (2012): 1251–77.

11. Malone et al., "Variation in the Social Systems of Extant Hominoids." One might argue that these characteristics are only present in all apes due to independent innovation along each distinct evolutionary lineage. While possible, it is highly unlikely given the overlap in these characteristics and not others.

12. Cf. Andrew Whiten and David Erdal, "The Human Socio-Cognitive Niche and its Evolutionary Origins," *Philosophical Transactions of the Royal Society B* 367 (2012): 2119–29.

13. M. Enquist, S. Ghirlanda, A. Jarrick and C. A. Wachtmeister, "Why Does Human Culture Increase Exponentially?" *Theoretical Population Biology* 74 (2008): 46–55; Kim Sterelny, *The Evolved Apprentice: How Evolution Made Humans Unique* (Cambridge, MA: MIT Press, 2012).

14. Arbib, "From Mirror Neurons to Complex Imitation."

15. Scott-Phillips, "Nonhuman Primate Communication."

16. See Atsushi Iriki and Miki Taoka, "Triadic (Ecological, Neural, Cognitive) Niche Construction: A Scenario of Human Brain Evolution Extrapolating Tool Use and Language from the Control of Reaching Actions," *Philosophical Transactions of the Royal Society B* 367 (2012): 10–23; Dietrich Stout and Nada Khreisheh, "Skill Learning and Human Brain Evolution: An Experimental Approach," *Cambridge Archaeological Journal* 25.4 (2015): 867–75.

17. Michael Tomasello, *The Cultural Origins of Human Cognition* (Cambridge, MA: Harvard University Press, 1999); Michael Tomasello, *The Natural History of Human Thinking* (Cambridge, MA: Harvard University Press, 2014).

18. Fuentes, "Integrative Anthropology and the Human Niche."

19. See Enquist et al., "Why Does Human Culture Increase Exponentially?"

20. Fuentes, "Integrative Anthropology and the Human Niche."

21. See Clive Gamble, John Gowlett, and Robin Dunbar, "The Social Brain and the Shape of the Paleolithic," *Cambridge Archaeological Journal* 21:1 (2011): 115–36; Kim Sterelny, *The Evolved Apprentice: How Evolution Made Humans Unique* (Cambridge, MA: MIT Press, 2012); and Fuentes, "Integrative Anthropology and the Human Niche."

22. Whiten and Erdal, "The Human Socio-Cognitive Niche."

23. Arbib, "From Mirror Neurons to Complex Imitation."

24. See Robin Dunbar, "The Social Brain: Mind, Language, and Society in Evolutionary Perspective," *Annual Review of Anthropology* 32 (2003): 163–81; Robin Dunbar, Clive Gamble, and John Gowlett, *Social Brain, Distributed Mind* (Oxford: Oxford University Press, 2010); and Gamble et al., "The Social Brain."

25. See Jeremy Kendal, "Cultural Niche Construction and Human Learning Environments: Investigating Sociocultural Perspectives," *Biological Theory* 6.3 (2012): 241–50; Fuentes, "Integrative Anthropology and the Human Niche."

26. However, see Arbib, "From Mirror Neurons to Complex Imitation," and Scott-Phillips, "Nonhuman Primate Communication."

27. See Iriki and Taoka, "Triadic Niche Construction," and Stout and Khreisheh, "Skill Learning and Human Brain Evolution."

28. Lyn Wadley, "Recognizing Complex Cognition through Innovative Technology in Stone Age and Palaeolithic Sites," *Cambridge Archaeological Journal* 23 (2013): 163–83; Marc Kissel and Agustín Fuentes, "From Hominid to Human: The Role of Human Wisdom and Distinctiveness in the Evolution of Modern Humans," *Philosophy, Theology and the Sciences* 3.2 (2016): 217–44.

29. See Arbib, "From Mirror Neurons to Complex Imitation"; Iriki and Taoka, "Triadic Niche Construction"; and Dietrich Stout and Thierry Chaminade, "Stone Tools, Language and the Brain in Human Evolution," *Philosophical Transactions of the Royal Society B* 367 (2012): 75–87.

30. See Matt Grove and Fiona Coward, "From Individual Neurons to Social Brains," *Cambridge Archaeological Journal* 18 (2008): 387–400; Sterelny, *The Evolved Apprentice*; and Stout and Chaminade, "Stone Tools, Language and the Brain." Here "broad bandwidth" is a term deployed by Kim Sterelny to imply the capacity to share information-rich and diverse content between individuals in relatively cohesive formats. This is not "language" per se, but more than simply multiple layers of code-model information transfer.

31. Sterelny, *The Evolved Apprentice*.

32. Ibid.

33. Arbib, "From Mirror Neurons to Complex Imitation."

34. Ibid., 263. Here Arbib refers to novel restructuring of neurological connectivity combined with changes to the throat and mouth musculature and ligamentary systems.

35. Ibid.; Tomasello, *The Natural History of Human Thinking*.

36. Stephen Mithen, "Overview and Response to Reviewers of The Singing Neanderthals," *Cambridge Archaeological Journal* 16.1 (2006): 97–112; *The Singing Neanderthals: The Origins of Music, Language, Mind and Body* (Cambridge, MA: Harvard University Press, 2007).

37. Mithen, "Overview."

38. Maxine Sheets-Johnstone, *The Primacy of Movement* (Amsterdam: John Benjamins, 1998).

39. Merlin Donald, *A Mind So Rare: The Evolution of Human Consciousness* (New York: Norton, 2001).

40. In contemporary theological and philosophical contexts one might expect to see René Girard cited here in a discussion of mimesis, but Girard's use of mimesis and "mimetic desire" is fairly different from that of Donald or Maxine Sheets-Johnstone, and not as directly related to the use of the concept in this context.

41. Kissel and Fuentes, "From Hominid to Human"; cf. Kissel, this volume.

42. Scott-Phillips, "Nonhuman Primate Communication," 60.

43. See, for example, Celia Deane-Drummond, *The Wisdom of the Liminal: Evolution and Other Animals in Human Becoming* (Grand Rapids, MI: William B. Eerdmans, 2014).

BIBLIOGRAPHY

Arbib, Michael A. "From Mirror Neurons to Complex Imitation in the Evolution of Language and Tool Use." *Annual Review of Anthropology* 40 (2011): 257–73.

Deacon, Terrence. *The Symbolic Species*. London: Penguin Press, 1997.

Deane-Drummond, Celia. *The Wisdom of the Liminal: Evolution and Other Animals in Human Becoming*. Grand Rapids, MI: Eerdmans, 2014.

Donald, Merlin. *A Mind So Rare: The Evolution of Human Consciousness*. New York: Norton, 2001.

Dunbar, Robin. "The Social Brain: Mind, Language, and Society in Evolutionary Perspective." *Annual Review of Anthropology* 32 (2003): 163–81.

Dunbar, Robin, Clive Gamble and John Gowlett. *Social Brain, Distributed Mind*. Oxford: Oxford University Press, 2010.

Enquist, M., S. Ghirlanda, A. Jarrick and C. A. Wachtmeister. "Why Does Human Culture Increase Exponentially?" *Theoretical Population Biology* 74 (2008): 46–55.

Foley, Robert, and Clive Gamble. "The Ecology of Social Transitions in Human Evolution." *Philosophical Transactions of the Royal Society B* 364 (2009): 3267–79.

Fuentes, Agustín. "Integrative Anthropology and the Human Niche: Toward a Contemporary Approach to Human Evolution." *American Anthropologist* 117.2 (2015): 302–15.

Gallese, Vitorio, and George Lakoff. "The Brain's Concepts: The Role of the Sensorimotor System in Conceptual Knowledge." *Cognitive Neuropsychology* 21 (2005): 455–79.

Gamble, Clive, John Gowlett, and Robin Dunbar. "The Social Brain and the Shape of the Paleolithic." *Cambridge Archaeological Journal* 21.1 (2011): 115–36.

Grove, Matt, and Fiona Coward. "From Individual Neurons to Social Brains." *Cambridge Archaeological Journal* 18 (2008): 387–400.

Iriki, Atsushi, and Miki Taoka. "Triadic (Ecological, Neural, Cognitive) Niche Construction: A Scenario of Human Brain Evolution Extrapolating Tool Use and Language from the Control of Reaching Actions." *Philosophical Transactions of the Royal Society B* 367 (2012): 10–23.

Kendal, Jeremy. "Cultural Niche Construction and Human Learning Environments: Investigating Sociocultural Perspectives." *Biological Theory* 6:3 (2012): 241–50.

Kissel, Marc, and Agustín Fuentes. "From Hominid to Human: The Role of Human Wisdom and Distinctiveness in the Evolution of Modern Humans." *Philosophy, Theology and the Sciences* 3.2 (2016): 217–44.

Malone, N. M., A. Fuentes, and F. J. White. "Variation in the Social Systems of Extant Hominoids: Comparative Insight into the Social Behaviour of Early Hominins." *International Journal of Primatology* 33.6 (2012): 1251–77.

Mithen, Stephen. "Overview and Response to Reviewers of The Singing Neanderthals." *Cambridge Archaeological Journal* 16.1 (2006): 97–112.

———. *The Singing Neanderthals: The Origins of Music, Language, Mind and Body*. Cambridge, MA: Harvard University Press, 2007.

Scott-Phillips, Thomas C. "Nonhuman Primate Communication, Pragmatics, and the Origins of Language." *Current Anthropology* 56.1 (2015): 56–80.

Seyfarth, R. M., D. L. Cheney, and P. Marler. "Monkey Responses to Three Different Alarm Calls: Evidence of Predator Classification and Semantic Communication." *Science* 210 (1980): 801–3.

Sheets-Johnstone, Maxine. *The Primacy of Movement*. Amsterdam: John Benjamins, 1998.

Sterelny, Kim. *The Evolved Apprentice: How Evolution Made Humans Unique*. Cambridge, MA: MIT Press, 2012.

Stout, Dietrich, and Nada Khreisheh. "Skill Learning and Human Brain Evolution: An Experimental Approach." *Cambridge Archaeological Journal* 25:4 (2015): 867–75.

Stout, Dietrich, and Thierry Chaminade. "Stone Tools, Language and the Brain in Human Evolution." *Philosophical Transactions of the Royal Society B* 367 (2012): 75–87.

Tomasello, Michael. *The Cultural Origins of Human Cognition*. Cambridge, MA: Harvard University Press, 1999.

———. *The Natural History of Human Thinking*. Cambridge, MA: Harvard University Press, 2014.

———. *Origins of Human Communication*. Cambridge, MA: MIT Press, 2008.

Tomasello, Michael, and Josep Call. "The Role of Humans in the Cognitive Development of Apes Revisited." *Animal Cognition* 7.4 (2004): 213–15.

Wadley, Lyn. "Recognizing Complex Cognition through Innovative Technology in Stone Age and Palaeolithic Sites." *Cambridge Archaeological Journal* 23 (2013): 163–83.

Whiten, Andrew, and David Erdal. "The Human Socio-Cognitive Niche and its Evolutionary Origins." *Philosophical Transactions of the Royal Society B* 367 (2012): 2119–29.

Index

About the Contributors

Marcus Baynes-Rock is a postdoctoral research fellow at the University of Notre Dame. He has long held an interest in relationships between humans and other animals, whether in evolutionary history or in contemporary societies. This has taken him to some markedly different fieldsites. In Ethiopia, he studied relations between people and spotted hyenas in the city of Harar, shared identity among Oromo farmers and their horses, and conflict and co-existence among Oromo pastoralists and spotted hyenas and leopards. In Australia, Marcus investigated farmers' conceptions of relations between wild canids and their livestock animals. He currently researches new forms of farming in Australia where crocodiles, emus, and stingless bees are undergoing varying degrees of domestication.

Dylan Belton is a PhD candidate in Systematic Theology at the University of Notre Dame and a John Templeton Foundation Graduate Student Scholar at the Center for Theology, Science and Human Flourishing. His research focuses on the relation between theology, philosophy, and the modern natural sciences, especially evolutionary biology. In his dissertation research, he aims to trace the relation between the philosophy of the organic and theology within the Darwinian era.

Ben Campbell is a senior lecturer in the Department of Anthropology at Durham University (UK). He has worked since the late 1980s with the Tamang-speaking communities of northern Nepal. He has twice been funded by the UK Economic and Social Research Council for research on indigenous knowledge and environmental practices, and has published on agricultural change, exchange labor in village subsistence, environmental policy, politics of sustainability, ritual community, pilgrimage and conversion. He has made the film *The Way of the Road* about turning a "remote" valley into a hub of globalization. Recent work explores renewable energy transitions for the global south.

Stewart Clem is a PhD candidate in Theology at the University of Notre Dame and a John Templeton Foundation Graduate Student Scholar at the Center for Theology, Science and Human Flourishing. He holds an M.Div. from Duke University and an MA and BA from Oklahoma State University. His research interests include the moral theology of Thomas

Aquinas, the ethics of lying and truth-telling, and issues surrounding free speech and public policy. His articles have appeared in journals including *Religious Studies, New Blackfriars,* and the *Journal of the Society of Christian Ethics*. He is a priest in the Episcopal Church and lives with his family in South Bend, Indiana.

Celia Deane-Drummond is professor of Theology at the University of Notre Dame and director of the Center for Theology, Science and Human Flourishing. She holds two honors degrees in natural science and theology and two doctorates (plant physiology and systematic theology). Her research interests are in the engagement of systematic and moral theology and the biological sciences, including specifically ecology, evolution, genetics, animal behavior, psychology, and evolutionary anthropology. She is joint editor of the international journal *Philosophy, Theology and the Sciences*. She has been chair of the European Forum for the Study of Religion and Environment since 2011 and is honorary visiting professor in Theology and Science at the University of Durham. She has published over two hundred scholarly theology and scientific articles or book chapters and twenty-five books as editor or author. A selection of her more recent books includes *Wonder and Wisdom: Conversations in Science, Spirituality and Theology* (2006); *Ecotheology* (2008); *Christ and Evolution* (2009); *Creaturely Theology,* with David Clough (2009); *Religion and Ecology in the Public Sphere,* with Heinrich Bedford-Strohm (2011); *Animals as Religious Subjects,* with Rebecca Artinian-Kaiser and David Clough (2013); *The Wisdom of the Liminal: Human Nature, Evolution and Other Animals* (2014); *Re-Imaging the Divine Image* (2014); *Technofutures, Nature and the Sacred,* with Sigurd Bergmann and Bronislaw Szerszynski (2015); *Ecology in Jürgen Moltmann's Theology,* 2nd edition (2016); and *Religion in the Anthropocene,* with Sigurd Bergmann and Markus Vogt (2017).

Julia Feder is assistant professor of Theology at Creighton University in Omaha, Nebraska. She specializes in systematic theology, with a particular focus on theological anthropology, suffering, and salvation. Her forthcoming book, *Trauma and Salvation: A Theology of Healing,* draws on the writings of Spanish mystic Teresa of Avila and Flemish political theologian Edward Schillebeeckx to construct healing from sexual trauma as a mystical-political practice. Before joining the faculty at Creighton, she was a postdoctoral research fellow at the Center for Theology, Science and Human Flourishing at the University of Notre Dame, where she examined the centrality of symbol-making in the human niche. She continues to think and write theologically about human evolution, drawing connections between Schillebeeckx's understanding of human experience and imagination (particularly the role of "negative contrast experiences") and biological anthropologists' understanding of human symbolic behavior and cognition. Her articles appear in *Philosophy, Theology, and the Sciences* and *Theological Studies*.

Agustín Fuentes completed a BA in Zoology and Anthropology, and an MA and PhD in Anthropology at the University of California, Berkeley, and is a professor and chair of Anthropology at the University of Notre Dame. His current research focuses on cooperation, community, and semiosis in human evolution, ethnoprimatology and multispecies anthropology, evolutionary theory, and interdisciplinary approaches to human nature(s). Fuentes' recent books include *Race, Monogamy, and Other Lies They Told You: Busting Myths about Human Nature* (University of California); *Conversations on Human Nature(s)* (with Aku Visala, Routledge); and *The Creative Spark: How Imagination Made Humans Exceptional* (Dutton).

Craig Iffland is a doctoral candidate in Moral Theology at the University of Notre Dame and a John Templeton Foundation Graduate Student Scholar at the Center for Theology, Science and Human Flourishing. His dissertation is entitled "Following and Not-Following the Divine Law," and explores the understanding of "sin" as a breach of divine law. In 2015–2016, he served as a visiting research fellow in International Law and Ethics at the Afro-Middle East Centre in Johannesburg, South Africa, where he studied the targeting policies of the United States in the "war on terror." He holds master's degrees in theology and philosophy from the University of Notre Dame and the University of St. Andrews, respectively.

Jeremy Kendal is a senior lecturer in the Department of Anthropology at Durham University. He received a PhD from Cambridge University (Sub-Department of Animal Behaviour), before completing postdoctoral positions at Stanford University with Marc Feldman and the University of St. Andrews with Kevin Laland. He took up a UK Research Council fellowship in the Department of Anthropology at Durham University, where he has remained. His research originated in biology and still largely uses quantitative techniques to investigate cultural dynamics and variation, but is now fueled by both the evolutionary and socio-cultural anthropological scholarship at Durham, where he directs the Centre for the Coevolution of Biology and Culture. His research interests include the effects of learning processes in socially and culturally constructed environments on the evolution of cultural variation, and a related interest modelling the cultural transmission of health-affecting behaviors. Current projects include studies with both human and non-human animals.

Marc Kissel is lecturer in the Department of Anthropology at Appalachian State University. He completed his PhD at the University of Wisconsin-Madison, where he studied how genetic models of human evolution fit with anthropological data. He was a postdoctoral research fellow in the Department of Anthropology at the University of Notre Dame, where he worked on a project on the evolution of human symbolic thought. His research includes the study of modern human origins, the evolutionary arc of human warfare, the application of quantitative genetics to anthropological data sets, and the causes and consequences of modern human variation.

Adam Willows is a postdoctoral research fellow at the Center for Theology, Science and Human Flourishing at the University of Notre Dame. He specializes in philosophical theology, with particular interests in normative and meta-ethics, virtue, free will theory, and philosophy of religion. His work emphasizes the potential for philosophical and theological traditions to interact, often to mutual benefit. His doctoral thesis, "A Defence of Theological Virtue Ethics," is a study of how theological thought on the virtues is well placed to respond to philosophical criticisms of virtue ethics. At Notre Dame, he studies the relationship between philosophical theology and the sciences, with special reference to moral theory and the underlying metaphysical commitments that shape interdisciplinary interaction. He has also produced work on moral transhumanism and is currently engaged in a study of moral exemplarism.

www.ingramcontent.com/pod-product-compliance
Ingram Content Group UK Ltd.
Pitfield, Milton Keynes, MK11 3LW, UK
UKHW020212290926
15470UKWH00001B/149

9 781498 548458